Arguing Science

Arguing Science

A Dialogue on the Future of Science and Spirit

Rupert Sheldrake
&
Michael Shermer

Monkfish Book Publishing Company
Rhinebeck, New York

Arguing Science: A Dialogue on the Future of Science and Spirit © 2016 by Rupert Sheldrake, Michael Shermer and TheBestSchools.org

All rights reserved. No part of this book may be used or reproduced in any manner without written permission from the publisher except in critical articles and reviews. Contact the publisher for information.

Printed in the United States of America.

Paperback ISBN: 978-1-939681-57-7
eBook ISBN: 978-1-939681-58-4

Library of Congress Cataloging-in-Publication Data
Names: Sheldrake, Rupert. | Shermer, Michael.
Title: Arguing science : a dialogue on the future of science and spirit / Rupert Sheldrake and Michael Shermer.
Description: Rhinebeck, New York : Monkfish Book Publishing Company, 2016.
Identifiers: LCCN 2016025408 | ISBN 9781939681577 (alk. paper)
Subjects: LCSH: Philosophy and science. | Science--Philosophy. | Physics--Philosophy. | Philosophy of nature.
Classification: LCC Q175 .S5328 2016 | DDC 501--dc23
LC record available at https://lccn.loc.gov/2016025408

Monkfish Book Publishing Company
22 East Market Street
Suite 304
Rhinebeck, New York 12572
USA (845) 876-4861
www.monkfishpublishing.com

TABLE OF CONTENTS

PREFACE

In 2015, TheBestSchools.org website hosted an intensive Dialogue on the Nature of Science between two well-known science writers, Rupert Sheldrake and Michael Shermer. Those dialogues form the basis of this book. While the book's editors have seen fit to rearrange some of the sequencing of the dialogue (as it appears online) and have made minor modifications in the language for the sake of clarity and readability, the dialogue published here is, in all essential matters, the same.

Drs. Sheldrake and Shermer agreed to take part in this dialogue over three consecutive months, with each month devoted to a particular topic. In the first month, May, the focus was on materialism in science. Dr. Sheldrake argued that science needs to free itself from materialist dogma; indeed, science misunderstands nature by being wedded to purely materialist explanations. Dr. Shermer opposed Dr. Sheldrake's position, arguing that science, properly conceived, is a materialistic enterprise; for science to look beyond materialist explanations is to betray science and engage in superstition. In June, the focus was on mental action at a distance. Dr. Sheldrake defended the position that telepathy, ESP, and psychic/psi phenomena in general are real and backed up by convincing evidence; their investigation deserves to be part of science. Dr. Shermer opposed Dr. Sheldrake's position, arguing that psychic or psi phenomena are artifacts of poor experimental procedure or outright fraud; no convincing evidence or experiments support their reality. During the third month, July, the focus was on God and science. Dr. Sheldrake put forth the notion that there is no conflict between science and the existence of God and that evidence from conscious experience renders belief in God reasonable. Dr. Shermer opposed Dr. Sheldrake's position, arguing that science in no way supports the existence of God; in fact, science undercuts the reasonableness of belief in God.

In order to provide a suitable context for this dialogue, Drs. Sheldrake and Shermer agreed to be interviewed by The Best Schools prior to their dialogue. It is our hope that you, the reader, will enjoy this feisty and thoughtful contest of ideas as much as we have enjoyed the process of publishing them.

THEBESTSCHOOLS.ORG
RUPERT SHELDRAKE INTERVIEW

Rupert Sheldrake, a Cambridge-trained biochemist and plant physiologist, is a prominent public intellectual critical of the authoritarianism and closed-mindedness that he finds increasingly typical of mainstream science. Sheldrake is the author of numerous bestselling books, including *A New Science of Life* (1981), *The Rebirth of Nature* (1990), *The Presence of the Past* (1988), *Dogs That Know When Their Owners Are Coming Home* (1999), *The Sense of Being Stared At* (2003), and, most recently, *Science Set Free: 10 Paths to New Discovery* (2012). This interview will focus especially on *Science Set Free* (titled *The Science Delusion* in the U.K.), which concentrates on the scientific enterprise as such and the obstacles to its proper pursuit.

Sheldrake has taken on the role of scientific "gadfly" in the proud tradition of Socrates, urging scientists to question received wisdom and to remove ideological blinders.

James Barham for TheBestSchools.org: Thank you very much for agreeing to this interview. Would you begin by giving us a quick sketch of your background? When and where were you born? What were your family's circumstances? What was your religious upbringing, if any? And please describe your education and early career.

Rupert Sheldrake: I was born on June 28, 1942, in Newark-on-Trent, Nottinghamshire, in the English Midlands, and was brought up there. My family was devoutly Methodist. I went to a high church Anglican boarding school. I was for a while torn between these two very different traditions—one Protestant and the other Anglo-Catholic with incense and all the trappings of Catholicism.

From a very early age I was interested in plants and animals. My father was an amateur naturalist and pharmacist and he encouraged this interest. My mother put up with it. I kept lots of animals at home.

I knew from quite an early age that I wanted to do biology, and I specialized in science at school. Then I went to Cambridge, where I studied biology and biochemistry. However, as I proceeded in my studies, a great gulf opened between my original inspiration— namely, an interest in actual living organisms—and the kind of biology I was taught: orthodox, mechanistic biology, which essentially denies the life of organisms, but instead treats them as machines.

There seemed to be very little connection between the direct experience of animals and plants and the way I was learning about them, manipulating them, dissecting them into smaller and smaller bits, getting down to the molecular level, and seeing them as evolving by blind chance and the blind forces of natural selection.

I felt more and more that there was something wrong, but I couldn't put my finger on it. No one else seemed to think there was anything wrong. Then a friend who was studying literature lent me a book on German philosophy containing an essay on the writings of Goethe, the poet and botanist.[1]

I discovered that Goethe, at the beginning of the 19th century, had a vision of a different kind of science—a holistic science that integrated direct experience and understanding. It didn't involve breaking everything down into pieces and denying the evidence of one's senses. This discovery—the idea that there could be a different kind of natural science—filled me with great excitement. So invigorated was I by this prospect that I wanted to find out why science had become so mechanistic. I was fortunate to get a fellowship at Harvard, where I spent a year studying the history and philosophy of science. Thomas Kuhn's book *The Structure of Scientific Revolutions*[2] had recently come out, and it had a big influence on me, gave me a new perspective. It made me realize that the mechanistic theory of life was what Kuhn called a "paradigm"—a collectively held model of reality, a belief system. He showed that periods of revolutionary change involved the replacement of old scientific paradigms by new ones. If science had changed radically in the past, then perhaps it could change again in the future. I was very excited by that.

When I got back to Cambridge [England], I did a Ph.D. on how plants develop, particularly working on the hormones within plants. I went on with my research on plant development and became a research fellow of Clare College in Cambridge and also a research fellow of the Royal Society, which gave me tremendous freedom, for which I'm very grateful.

James Barham: Spirituality, mysticism, and Christianity have played a prominent role in your life. From your writings, one gets the sense that you are on a pilgrimage or quest. Could you elaborate on this aspect of your life? Was there a point in your life where you experienced what might be called a "religious conversion" (or perhaps several)?

Rupert Sheldrake: When I received a grant in 1968 from the Royal Society to go and study tropical plants in Malaysia, at the University of Malaya, I traveled through India on the way there. I found India a very exciting place to be, and as I traveled through that country I encountered gurus and ashrams and temples, which opened my eyes to a range of phenomena I was completely unfamiliar with.

When I got back to England I got interested in exploring consciousness, and I had various psychedelic experiences, which convinced me that the mind was vastly greater than anything I'd been told about in my scientific education. Then I got interested in transcendental meditation because I wanted to be able to explore consciousness without drugs. I was increasingly intrigued by India, by yoga, and by meditation, and in 1974 I had a chance to go and work in India as principal plant physiologist at the International Crops Research Institute for the Semi-Arid Tropics (ICRISAT) in Hyderabad.

I was thrilled by the idea of immersing myself in this exotic and fascinating culture. While I was in India, I visited temples and ashrams and I attended discourses by gurus and holy men. I also took up Sufism, and had a Sufi teacher in Hyderabad, who was the grandfather of a friend of mine. He gave me a Sufi mantra, a *wazifa*, which for about a year I practiced in a Sufi form of meditation. But I didn't want to become a Sufi because in India to become a Sufi, you have to be a Muslim first and foremost, and that would have been too much of a stretch.

Then, an original idea crossed my mind: What about the Christian tradition? I hadn't given it a thought. I spoke to a Hindu guru, and he said, "All paths lead to God. You come from a Christian family so you should follow a Christian path."

The more I thought about it, the more sense it made, and I began to pray with the Lord's Prayer, and I started going to church at the Anglican Church, St. John's, Secunderabad. After a while I was confirmed, at the age of 34, by an Indian bishop in the Church of South India (an ecumenical church formed by the coming together of Anglicans and Methodists). I felt very happy to be reconnected with the Christian tradition.

I still felt a huge tension between the Hindu wisdom, which I felt was so deep, and the Christian tradition that seemed a bit shallow on the spirituality side. I then discovered a wonderful teacher, Father Bede Griffiths, who had a Christian ashram in South India. He was an English Benedictine monk who had lived there for 25 years when I met him. His ashram combined many aspects of Indian culture with Christian tradition. I wrote my first book, *A New Science of Life*, in his ashram.

When I got back to England, after a long period in India, I had a wonderful time rediscovering the English tradition. I rediscovered sacred places—England is full of them, great cathedrals and churches—and I started going to my local parish church in Newark-on-Trent and to cathedrals, where there is marvelous singing. Instead of just seeing it as an aesthetic experience as I had done before, I now felt part of it and was very, very moved by it and felt privileged to be part of this tradition. So, since then it's been my practice to go to church on Sundays whenever I can. I see the creeds first and foremost as statements of belief in God's threefold nature. The doctrine of the Holy Trinity makes great sense to me. No doubt I differ from some people in my interpretation of the details. But probably even the most unbending literalists do not accept every part of it without some qualifications. For example, in the Apostle's Creed when it says of Jesus Christ that "he sitteth on the right hand of God the Father Almighty," is he really sitting? And are God the Father and Jesus located in a particular place, a celestial throne-room? And does God the Father have right and left hands?

James Barham: To what extent do you think the maverick path you ended up taking was forced on you by the empirical data themselves, and to what extent was it contingent on your being exposed to alternative ways of thinking in India and elsewhere? In other words, do you think you would have become the Rupert Sheldrake of today without the experience of living and working in [at least two] radically different cultural environments?

Rupert Sheldrake: When I was 17, in the gap between leaving school and going to study at Cambridge, I worked as a temporary lab technician in a pharmaceutical laboratory because I wanted to get some research experience. What I didn't know when I took the job was that it was a vivisection facility. Working there made me ask some deep questions about animals, animal suffering, scientific objectivity, and mechanistic attitudes to nature, which were put into practice on a daily basis in this laboratory, which was a kind of death camp for animals.

By the time I was studying biology as an undergraduate at Cambridge, I was already very doubtful about the reductionist and mechanist approach, which is why I welcomed the opportunity to study the history and philosophy of science at Harvard—to get a bigger perspective.

After Harvard, when I returned to Cambridge [England] in 1964 and was doing research on plant development, I became convinced the molecular and reductionist approach would never enable us to understand the development of form. I became interested in the morphogenetic field concept, first put forward in the 1920s.[3] Although I traveled in India and lived in Malaysia, it was reading books on theoretical biology and philosophy—especially the philosophy of Henri Bergson in his book *Matter and Memory*[4]—that led to my developing the hypothesis of morphic resonance. This was in 1973, while I was still in Cambridge, before I went to live and work in India in agricultural research. I continued to develop these ideas, and India was a good place to do this. After all, in Hindu philosophy the idea of a kind of memory in nature is commonplace, as it is in Buddhist philosophy.

My idea of an inherent memory in nature through morphic resonance did not seem weird to most of my Indian colleagues and friends. India provided a friendly environment for writing my first book, *A New Science of Life*.[5] But the basic ideas came from Western science and Western philosophy.

James Barham: In your new book, *Science Set Free*[6], you speak of the "intellectual phase-locking"—that is, the "group think" or "herd mentality"—that clearly plagues mainstream science today. We were wondering whether this was mainly due to the hubris that comes from the unprecedented social prestige scientists now enjoy, or whether it might not be more a matter of the metaphysical commitment to materialism that has been deeply ingrained in the scientific community for the past 400 years.

In other words, is the intellectual phase-locking of scientists more about arrogance and turf-protecting? Or is it more about their being in the grip of a misguided ideology? Or both? Please elaborate.

Rupert Sheldrake: The materialist ideology promotes a high degree of conformity in scientific thinking because it is indeed ideological, and materialists are unforgiving towards heretical deviations from this belief system. Over the course of the 20th century, the atmosphere within biology became increasingly intolerant at the same time as physics opened up a wider range of possibilities. There are still great limitations on what professional physicists can think, but there is a toleration of alternative interpretations of quantum mechanics, divergent interpretations of cosmology, the question of whether there is one universe or many, and so on.

Another reason for the greater uniformity of thinking is the professionalization of science. In the 19th century, many of the most creative scientists were not professionals. For example, Charles Darwin was an amateur naturalist living on a private income, with no academic post or government grant. He was much freer as a result. Now, the vast majority of scientists rely on salaries and are far more aware of peer-group pressure. In fact, the peer-review system for jobs, grant applications, and publication of papers in journals means that peer pressure dominates their lives. In the nineteenth century, there were fewer constraints on creative and independent thinking.

James Barham: Like you, we at TBS are very much interested in doing what we can to help "extend the boundaries of what is not regarded as unthinkable," as Thomas Nagel put it in his recent book, *Mind and Cosmos*[7]. The reasons are many, but the overriding one is the danger we believe *scientism*[8] poses to human freedom and dignity, as well as to morality and limited self-government.

At the same time, we believe that the most obvious way to reform science is by demonstrating a better way forward that is recognizable as such to scientists themselves. In other words, give scientists a better way of doing science and let them vote with their feet.

Accordingly, we would like to devote a good part of this interview to pressing you on a number of scientific points, teasing apart what seem to us to be your most promising hypotheses and speculations, using your new book, *Science Set Free*, as a constant point of reference. So, here goes.

First, let's discuss "morphic resonance," which appears to be your most widely discussed contribution. Could you begin by giving our readers a thumbnail sketch of the theory?

Rupert Sheldrake: In brief, morphic resonance is the hypothesis that there is a kind of inherent memory in nature. In the most general terms, the "laws of nature" are more like habits. Within each species, each individual draws upon a collective memory and in turn contributes to it. My proposal is that this works on the basis of similarity: the similarity of three-dimensional vibratory patterns in self-organizing systems. Self-organizing systems include atoms, molecules, cells, organs, organisms, societies of organisms (like flocks of birds), solar systems, and galaxies. This definition excludes systems that do not organize themselves, like tables, chairs, and machines, which are put together according to external designs, to serve external purposes.

James Barham: In a recent interview, you wrote: "The idea that animals and plants are machines is really Dogma Number One." To which we can only say: "Amen!" We, too, feel that it is in the arena of rethinking the fundamental nature of living systems—in "seeing past Darwin," as we like to put it—that the fight to defend the human spirit against scientism can be most effectively joined. And that is one reason why we are so interested in your morphic resonance hypothesis.

We see the Darwinian, reductionist approach to teleology, or goal-directedness—a property that is manifest in all living systems—as lying at the intellectual root of scientism, and we see nonlinear dynamics as a very fruitful way of tackling the problem of teleology head-on, by plowing straight through the Darwinian roadblock. Would you agree? Or does your interest in nonlinear dynamics lie in a different direction from ours?

Rupert Sheldrake: Dynamics is a branch of mathematical theory dealing with change, and a central concept in dynamics is that of the attractor. Instead of modelling what happens to a system by considering only the way it is pushed from behind, attractors in mathematical models provide an explanation in terms of a kind of pull from the future. The principal metaphor is that of a basin of attraction, like a large basin into which small balls are thrown. It would be very complicated to work out the trajectory of each individual ball starting from its initial velocity and angle at which it hit the basin; but a simpler way of modelling the system is to treat the bottom of the basin as an attractor: balls thrown in from any angle and at any speed will end up at the bottom of the basin.

In the 1950s, the British embryologist C.H. Waddington proposed that morphogenetic fields contained what he called "chreodes" (Greek for "necessary paths") which channeled developing organs and embryos towards attractors, understood as the form of the mature organ or organism. He compared organs developing under the influence of these fields to balls rolling down valleys.

Later, more technical mathematical models of morphogenetic fields and dynamical attractors were developed by René Thom and others. "Strange" or "chaotic" attractors, as they are called, are just one kind of attractor in dynamical systems theory.

The attractors within morphic fields are more complex, and perhaps less "chaotic." The word "morphic" comes from the Greek word *morphē*, meaning "form," and expresses the idea that morphic attractors pull developing systems towards them, and that the form of the attractor depends on a kind of memory given by morphic resonance.

Thus, for example, an oak seedling is attracted towards the mature form of an oak tree through the morphic attractor in its morphogenetic field. These attractors act as ends or goals, and in that sense are teleological, where "teleology" is the subject of ends or goals or purposes (from the Greek word *telos*, meaning "end" or "goal").

James Barham: Many would say that the whole point of the concept of a virtual, phase-space attractor is that it helps us conceive of teleological or goal-directed action in living systems in a way that does not require us to say that there is "backwards causation" of the future on the past. And yet you are not bashful about invoking backwards causation in your work. Why is that? Wouldn't it be preferable to avoid backwards causation, if possible?

Rupert Sheldrake: Attractors attract. In that sense, they imply teleology or final causation, or the pull of ends or goals. So, there is a kind of virtual backwards causation from virtual ends or goals. If I decide to visit San Francisco in six months' time, that acts as a kind of virtual attractor:

I book my airline tickets and make my arrangements in accordance with this plan, directed towards a future which does not yet exist. I am not saying that all this is caused by my future stay in San Francisco, because all sorts of unforeseen circumstances could prevent my actually getting there. Nevertheless, I think in some situations there is a kind of backwards causation.

This is one of the reasons that I take research within parapsychology seriously. I think there is good evidence for precognitive dreams, and also for presentiment, whereby an emotional arousal can have a physiological arousing effect five or six seconds in advance. Perhaps the intellectual world would be a neater place if we disregarded this evidence; but it can't be disregarded just because it does not fit into a particular theory of time and causation.

So, in summary, I think that ends or goals are given by virtual futures that pull organisms towards them, but sometimes there are influences from actual futures, rather like occasional memories of the future.

James Barham: Virtual attractors are purely mathematical concepts. The real question is: What type of physical field underlies the goal-directed behavior of living systems — including that of human beings — which then shows up as a closed, phase-space trajectory? Here, you speak of "morphic fields." Fair enough. But what is a morphic field, exactly?

Rupert Sheldrake: Fields are most generally defined as regions of influence. A magnetic field is within a magnet and is also a region of magnetic influence around it.

13

The earth's gravitational field is within the earth and stretches out invisibly far beyond it keeping the moon in its orbit.

Morphic fields are within and around self-organizing systems and contain attractors towards which the system develops. When a system has reached its final form—for example, in an insulin molecule or a Paramecium cell—its morphic field helps to stabilize its form and restore it after disturbances.

It's hard to say exactly what a morphic field is. There are mathematical theories, such as those of René Thom, but these are models in multi-dimensional phase space, which is a specialized mathematical concept utterly obscure to everyone except professional mathematicians.

So, does this tell us what the fields are? Not really. But what is the exact nature of an electromagnetic field? Electrical and magnetic fields were first proposed by Michael Faraday in the 1830s, and he was unsure as to their nature. He put forward two possibilities. First, that they consist of strains and patterns in subtle matter, called the "ether." Or, second, they were modifications of "mere space."

James Clerk Maxwell took up the ether hypothesis in the 1860s in his famous equations of electromagnetism, but Einstein dropped the idea of the ether and reverted to something closer to Faraday's idea of fields as modifications of mere space. Gravitational fields are also patterns in space. The gravitational field is not in space-time, according to Einstein, it is space-time.

Modern superstring theory tries to account for the physical fields of nature in terms of a 10- or 11-dimensional proto-field, in which the extra dimensions "curl up" to give the fields of nature as we know them.

So, what, exactly, is any field?

James Barham: You accept that psi phenomena (telepathy, ESP, paranormal activity, etc.) are real. What convinces you of their reality? And why do skeptics like James Randi, who make it their livelihood debunking psi phenomena, remain so unconvinced?

According to Dean Radin, psi phenomena have extremely strong statistical backing (with significance levels better than one in a billion), and yet they are weak in the sense that they don't permit anyone to beat the lottery or win consistently in Las Vegas. Do you agree?

Rupert Sheldrake: I think that phenomena like telepathy are real because they happen spontaneously in the course of normal life, and also they are supported by a great deal of experimental evidence. I have had many dealings with self-proclaimed skeptics and it has become obvious to me that their opposition to these phenomena is not based on a careful study of the evidence, but rather on materialist ideology, which says that minds are nothing but brains, and so if all mental activity is located inside the head, it cannot possibly have effects at a distance. Therefore, psychic phenomena like telepathy are impossible. And therefore all the evidence for them must be flawed or fraudulent, and people who believe in these things are subject to delusions.

In my various encounters with skeptics like Richard Dawkins, James Randi, Daniel Dennett, and Michael Shermer, I have found that they have no interest in looking at the evidence because they know in advance it must be false. In other words, their position is one of prejudice rather than open-minded scientific enquiry. In that sense, I think they are deeply anti-scientific.

Some of the phenomena studied under rather artificial conditions by parapsychologists show only fairly weak effects, but in the real world telepathy may operate much more reliably. For example, I have done studies on telepathy between mothers and their babies, and the mothers often know quite accurately when their baby needs them even when they are miles away. Similarly, many dogs and cats seem to know when their owners are coming home and wait for them at a door or window in a reliable and repeatable way.

I agree that psychic abilities may be much weaker when it comes to winning lotteries or beating the casino in Las Vegas, but these are not the kinds of situations in which psi is expressed in the real world. In my own research, I have concentrated on common, everyday psi phenomena that most people have personally experienced, like the sense of being stared at, pets knowing when their owners are coming home, telepathic bonds between mothers and children, and telephone telepathy (thinking of someone for no apparent reason who then calls).

James Barham: In Chapter 11 of *Science Set Free*, regarding the objectivity of science, you mention the very many ways in which scientists—who are, after all, only fallible human beings—may fall short of the ideal of objectivity. No one can quarrel with this. But then you make a pretty strong claim: "The supposed objectivity of the 'hard sciences' is an untested hypothesis."

We would like to say in reply that science is a normative enterprise, meaning that it can be done well or badly. The norms or ideals that govern—or ought to govern—scientific practice were summarized by Robert K. Merton long ago: disinterestedness; organized skepticism; transparency; and universality; among others. To which, of course, must be added the ordinary, everyday virtues of diligence, honesty, fairness, and so forth.

The question, then, is this: Are the authoritarianism and closed-mindedness that are increasingly plaguing science today the result of the inherent subjectivity of all human knowledge? Or are they not rather the result of the progressive breakdown of the traditional Mertonian norms due to new temptations (power, greed, ideology, fashion, etc.) that have arisen due to Big Science, and which are posing an unprecedented challenge to the ordinary fallen human nature of scientists?

In other words, is science really irremediably subjective? Or is it being corrupted, which implies it has fallen away from the ideal of objectivity?

Rupert Sheldrake: I am all in favor of the scientific ideals enunciated by Robert Merton and by others. But in science, as in any other human endeavor, there is an enormous gap between the ideals and reality. For example, Christians would presumably all subscribe to a belief in the importance of loving kindness and forgiveness, and yet many Christians have taken part in wars that involve mass slaughter and great cruelty. Everyone can see that there is a gulf between ideals and reality in ordinary life, without rejecting the ideals. But in the sciences, there has been a remarkable degree of self-deception through scientists believing their own rhetoric about objectivity. I'm not suggesting that most scientists are behaving fraudulently or deceitfully. But it has become increasingly apparent in the past few years that much of established science is a house of cards. For instance, recently a high proportion—more than 80 percent—of the key papers in biomedical science have turned out not to be replicable.

The same has become apparent in the realms of psychology and other sciences, too. The main reason for this unreliability seems to be that scientists publish only a small proportion of their data, usually the proportion that shows the most impressive results, namely, results that agree with their hypothesis. As much as 80 percent of the data may not be published because it does not fit in with the experimenters' expectations, or it does not make sense. This inevitably imparts a major bias to papers published in scientific journals.

For years, defenders of scientific orthodoxy have argued that the objectivity of science is guaranteed by replication and the peer-review process. But it has become glaringly obvious within the scientific world that this is not the case. First of all, scientists get very little credit for replicating other people's results. Such research is regarded as unoriginal and is generally discouraged. And even if scientists do carry out replications of other people's research, scientific journals will often refuse to publish them on the grounds they are not original. Journals also have a strong bias against publishing negative results.

Meanwhile, peer review is not necessarily a guarantee of quality. Moreover, it can militate against originality because peer reviewers, who operate anonymously, tend to defend the status quo. And many of them simply don't have time to read very thoroughly the papers they are asked to review. Recently, in an experiment on peer reviewing, dozens of nonsense papers, generated by computers, were submitted to peer-reviewed journals and more than half of them were accepted!

A new mood of humility is apparent within the sciences, and the complacency that for decades has enabled scientists to imagine that just because they were scientists they were objective is melting away. Discussion of questionable research practices is going on throughout the scientific world at present, and hopefully will lead to better research procedures.

James Barham: You end your new book with the wonderful line: "Much remains to be discovered and rediscovered, including wisdom." In your view, what is wisdom?

Rupert Sheldrake: Knowledge is about information, but wisdom is more about seeing patterns and the way in which things interact. It also involves taking a long-term perspective.

Unfortunately, in the modern world our perspectives are often very short-term, driven by daily news agendas, four- or five-year electoral cycles, and annual or quarterly profit reports. Investors in the stock market now make decisions on timescales of fractions of a second.

Wisdom involves looking at the bigger picture — taking a more holistic view — and it cannot easily be taught because in part it depends on experience, and often on intuition as well—a direct knowing that is not reducible to textbook facts or statistical procedures.

James Barham: Thank you very much for your time and your insights! Are there any final thoughts you would like to share with our readers? What changes in the scientific world would you like to see in the next five to 10 years? What is needed for these changes to be realized?

Rupert Sheldrake: As I argue in my book *Science Set Free*, I am convinced that the sciences are being imprisoned by the outmoded ideology of materialism. I show how each of the 10 dogmas of materialism can be turned into a question, treated as a scientific hypothesis, and evaluated scientifically. None of these dogmas turns out to be valid or persuasive. In every case, new questions open up, along with new possibilities for scientific research.

I would like to see these possibilities explored. There are already many open-minded scientists working within universities and other scientific institutions, but most of them are unable to follow unconventional lines of research because they're afraid these would not be funded. I would like to see a plurality of sources for funding in science that enable different approaches to be explored. This is unlikely to happen through government funding agencies, which are dominated by the science establishment, but there are many private foundations that could fund alternative scientific and medical research and I hope that some of them will do so.

I also hope that non-materialist scientists will feel able to meet up with other like-minded professionals and work together to change the sciences from within. And I hope that these open questions will become more widely known to students at schools through the educational system. For anyone interested in these possibilities, I recommend a new website, OpenSciences.org, that is a portal for the post-materialist sciences.

I am delighted that TBS is exploring these issues and hope that students in schools colleges and universities will be able to have some influence over what they are taught through making their interests known and through not blindly accepting the dogmas that are presented to them. Students need to learn, but they should also have some influence over what they are taught.

THEBESTSCHOOLS.ORG
MICHAEL SHERMER INTERVIEW

Michael Shermer is the editor-in-chief of *Skeptic* magazine and the author of several books, most recently *The Moral Arc*. He has also authored a dozen other books on science, evolution, religion, parapsychology, morality, and other topics, many of them bestsellers.

Dr. Shermer holds a B.A. in psychology/biology from Pepperdine University, an M.A. in experimental psychology from Cal State Fullerton, and a Ph.D. in the History of Science from Claremont Graduate University. Among his numerous endeavors, he has been writing the monthly "Skeptic" column for *Scientific American* magazine since 2001, has produced the 13-episode television series "Exploring the Unknown" for the Family Channel, and is a former competitive bicycle racer who co-founded Race Across America (RAAM) and helped design better protective equipment for the sport.

James Barham for TheBestSchools.org: Thank you very much for taking time out of your busy schedule to be interviewed. You have a new book just out! It is called *The Moral Arc: How Science and Reason Lead Humanity toward Truth, Justice, and Freedom*[1], and we will be spending a good deal of this interview discussing it in detail.

Before we do that, though, we would like you to tell us a little bit about yourself. First of all, when and where were you born, and what is your family's educational, social, ethnic, religious background, etc.?

Michael Shermer: I was born and raised in Southern California, specifically the La Canada area in the foothills surrounding Los Angeles. My parents were not religious and none of them went to college. I had both biological and step-parents, and toggled between homes weekdays and weekends while growing up—a real boon at Christmas time! I have three sisters and two brothers and am an only child. Figure that one out—the quintessential American blended family of two half-sisters (same father, different mother), a step-sister, and two step-brothers. No one in the family was particularly religious, and yet somehow we grew up learning moral principles and how to be good. Imagine that!

James Barham: Today, you are one of the most recognizable atheists/agnostics in the United States, as well as across the world. Yet, you were once an evangelical Christian. That's quite a journey! Could you describe the circumstances that led you to become an evangelical Christian as well as give some snapshots of what your life during that time was like? Is there anything you miss about that phase of your life?

Michael Shermer: My conversion to Christianity came at the behest of my best friend in high school, whose parents were Christian, and it was something of a "thing" to do at the time (early '70s) as the evangelical movement was just taking off. I accepted Jesus as my savior on a Saturday night with my friend, and the next day we attended the Glendale Presbyterian church, which had a very dynamic and histrionic preacher who inspired me to come forward at the end of the sermon to be saved. My buddy told me that I didn't need to do it, but it seemed more official in a church than at the bar at my parents' home. So, I was born again, again, so I figure that must count for something, you know, just in case I'm wrong now in my belief that there very probably is no God.

I took my religious beliefs fairly seriously. For a couple of years I attended this informal Christian study fellowship group at a place called "The Barn" in La Crescenta, which, in looking back, was a quintessential '70s-era hangout with a long-haired, hippie-type guitar-playing leader who read Bible passages that we discussed at length. But more than the social aspects of religion, I relished the theological debates, so I matriculated at Pepperdine University (a Church of Christ institution) with the intent of becoming a theologian. Although living in the Malibu hills overlooking the Pacific Ocean was a motivating factor in my choice of a college, the primary reason I went there was I thought I should attend a school where I would receive serious theological training, and I did.

I took courses in the Old and New Testaments, Jesus the Christ, and the writings of C.S. Lewis. I attended chapel twice a week (although, truth be told, it was required for all students). Dancing was not allowed on campus (the sexual suggestiveness might trigger already-inflamed hormone production to go into overdrive), and we were not allowed into the dorm rooms of members of the opposite sex. Despite the restrictions, it was a good experience; I was a serious believer and I thought this was the way we should behave.

The only thing I miss—and only a little—is the confident certainty that religion brings, the knowing absolutely that this is the One True Worldview. That was, as well, the downfall of my faith.

James Barham: To follow up on the last question, what circumstances led you to abandon evangelical Christianity? In repudiating evangelical Christianity, did you immediately become a skeptic of all religion, or did your skepticism evolve more gradually? Please explain.

Michael Shermer: While undertaking my studies at Pepperdine, I discovered that to be a professor of theology you needed a Ph.D., and such a doctorate required proficiency in Hebrew, Greek, Latin, and Aramaic. Knowing that foreign languages were not my strong suit (I struggled through two years of high school Spanish), I switched to psychology and mastered one of the languages of science: statistics.

In science, I discovered that there are ways to get at solutions to problems for which we can establish parameters to determine whether a hypothesis is probably right (like rejecting the null hypothesis at the 0.01 level of significance) or definitely wrong (not statistically significant). Instead of the rhetoric and disputation of theology, there was the logic and probabilities of science. What a difference this difference in thinking makes!

But the switch to science was only one factor in my deconversion. There was the intolerance generated by absolute morality, the logical outcome of knowing without doubt that you are right and everyone else is wrong. There were the inevitable hypocrisies that arise from preaching the *ought*, but practicing the *is*. One of my dormmates regularly prayed for sex, rationalizing that he could better witness for the Lord without all that pent-up libido. There was the awareness of other religious beliefs — often mutually exclusive — and their adherents, all of whom were equally adamant that theirs was the One True Religion. And there was the knowledge of the temporal, geographic, and cultural determiners of religious beliefs that made it obvious that God was made in our likeness and not the reverse.

By the end of my first year of a graduate program in experimental psychology at California State University, Fullerton, I had abandoned Christianity and stripped off my silver Ichthus medallion, replacing what was for me the stultifying dogmas of a 2,000-year-old religion with the worldview of an always-changing, always-fresh science. My enthusiasm for the passionate nature of this perspective was communicated to me most emphatically by my evolutionary biology professor, Bayard Brattstrom, particularly in his after-class discussions at a local bar — The 301 Club — that went late into the night. This was another factor in my road back from Damascus: I enjoyed the company and friendship of science people much more than that of religious people. Science is where the action was for me.

James Barham: You are known, among other things, as a skeptic, an agnostic, and an atheist. Is there a designation that you prefer for yourself? How would you distinguish these three designations?

Speaking for yourself, are you certain God does not exist? Some atheists have such an antipathy toward God that they might better be called anti-theists. You've never struck us as that hardcore. What accounts for that?

Michael Shermer: I'm an atheist. I don't believe in God. No, I am not 100 percent certain there is no God. But there is insufficient evidence to conclude that there is, and so pick whatever label you like. Technically speaking, "agnostic," as Thomas Huxley defined it in 1869 to mean that God is "unknowable," is accurate from an ontological perspective since it is difficult to imagine a scientific experiment that would clearly delineate between the God hypothesis and the no-God hypothesis. But we are behaving primates, not just thinking sapiens, so we must choose to act on our beliefs, and I act under the presumption that there is no God.

That said, I don't like to define myself by what I don't believe. I believe in lots of things: the Big Bang, evolution, the germ theory of disease, plate tectonics and the geological record, the laws of nature, and the like. I also believe in natural rights, moral progress, and that science and reason are the best tools we have for determining how best we should live. To that end, I call myself a humanist and I adopt the worldview of Enlightenment Humanism.

James Barham: In addition to being a best-selling book author and to writing a monthly "Skeptic" column for *Scientific American*, you are also the founder of the Skeptics Society and editor-in-chief of its house magazine, *Skeptic*. Could you tell us about the purpose of the Skeptics Society and how the idea for it came to you?

Michael Shermer: After I earned my Ph.D. in the history of science, I got a job teaching at Occidental College, a highly regarded four-year liberal arts college in Los Angeles, and I figured I would settle in for the duration. But I was still restless to be an entrepreneur, so I co-founded the Skeptics Society and *Skeptic* magazine in my garage as a hobby, and as it grew, I realized by the late '90s that I could do this full time. I loved teaching and being in a classroom. However, publishing magazines, writing a monthly column for such a large-circulation magazine as *Scientific American*, writing books, and doing television and radio shows gave me access to a much larger classroom than I could ever reach in a brick-and-mortar building. So, I've never looked back, even though I am now teaching one class a year at Chapman University — Skepticism 101 — a critical thinking course that I like to do to try out new ideas on students.

The mission of the Skeptics Society is to promote science and critical thinking. Although we do a lot of debunking—and let's face it, there's a lot of bunk out there—we always maintain an undercurrent of promoting the positive aspects of science, which we also do through our monthly science lecture series at Caltech and our annual conference on various topics.

James Barham: You have stated, in connection with non-mainstream scientific claims, that "Skepticism is the default position because the burden of proof is on the believer, not the skeptic."[2]

However, some of the people you have criticized in your *Scientific American* column and in *Skeptic* magazine — we are thinking especially of Rupert Sheldrake, with whom you will be engaging in a "Dialogue on the Nature of Science" here at TBS in the near future — have pointed out that they are the ones who are "skeptical" vis-à-vis mainstream scientific opinion.

In fact, there is now an entire website, Skeptical About Skeptics, devoted to equalizing the burden of proof between the scientific establishment and its critics.

How do you respond to Sheldrake and others who are "skeptical about your skepticism," and who want to shift the burden of proof back onto you?

Michael Shermer: My position on who has the burden of proof stands pretty solid among most scientists because of the fact that most mainstream scientific theories are hard won over many years and, like governments, "should not be changed for light and transient causes" (as Jefferson opined in the Declaration of Independence).

Yes, historically speaking, a few mainstream scientific theories were overturned by isolated outsiders, but that is almost never the case today. There's a reason we talk about a "consensus" among climate scientists that global warming is real and human-caused. It isn't because science depends on the consensus of authorities; it is because science is an extremely competitive enterprise, and if there were serious problems with climate models or datasets, then there is little doubt that these would have been uncovered by scientists working in other labs. The idea that scientists get together on weekends to get their story straight in the teeth of opposition from without is ludicrous. Attend scientific conferences on any topic and you will find often bitter contentions over this and that dataset or hypothesis. By the time findings and theories filter out of the lab into the public, they have been tried and tested and hold a high degree of confidence of most scientists who work in that field.

I will expand on this more in my dialogue with Rupert, but in short, there's certainly nothing wrong with outsiders (and especially insiders!) challenging the consensus. But the argument that they laughed at the Wright brothers doesn't hold because they laughed at the Marx brothers too, so being laughed at doesn't mean you're right. You have to actually have both data and theory in support.

James Barham: It is time to turn to your new book— by all accounts your magnum opus—*The Moral Arc: How Science and Reason Lead Humanity toward Truth, Justice, and Freedom.*

And a blockbuster of a book it is! First and foremost, it seems to us, *The Moral Arc* is a treasure chest of thought-provoking, cutting-edge social science research on a very wide array of topics, grouped around the unifying theme of the human condition very broadly conceived.

But in addition to its wealth of absorbing empirical detail, the book is also thesis-driven. The thesis—again in our interpretation—is twofold: (1) the moral progress of humanity over the past several centuries has been palpable, and may be confidently expected to continue into the future; and (2) the principal driver of that moral progress has been science and reason, with the corollary that religion has not only been of no help in this regard, but has been a positive hindrance—and therefore the sooner it is extirpated the better.

Is that a fair assessment of *The Moral Arc*, in very general terms?

Michael Shermer: Yes. *The Moral Arc* is by far my best and most important work, so thank you for recognizing that.

Most people have a hard time getting past the first thesis of the book—that things are getting better—and it is understandable why, if you're paying attention at all to the news with all the stories coming out of Syria, Ukraine, Iraq, and parts of Africa, not to mention Ferguson, Missouri, and Baltimore, Maryland! It seems like things are bad and getting worse. But we should follow the trend lines, not just the headlines, and when you do so, there is no question that in nearly every sphere of human endeavor, there has never been a better time to be alive than now.

As for my second thesis about religion, that is very much a secondary issue to the stronger thesis emphasizing the role of science and reason and the Enlightenment. *The Moral Arc* is not an "atheist" book. It's a science book. It is about the positive forces that have been at work over the past two centuries to expand the moral sphere—bend the moral arc—and grant more rights and freedoms and liberty and prosperity to more people in more places than at any time in history.

I don't care what someone's religion is, as long as they agree that everyone has the natural right to be treated equally under the law, to be endowed by nature and nature's laws—evolution in my model with inalienable rights to life, liberty, and the pursuit of happiness, and to honor the Liberty Principle: The freedom to think, believe, and act as we choose, so long as our thoughts, beliefs, and actions do not infringe on the equal freedom of others. I doubt that there are any Jews or Christians who would disagree with this principle, but then they—like me and most everyone else reading these words—are children of the Enlightenment, where these ideas were first articulated.

James Barham: We found ourselves—perhaps surprisingly—in general agreement with much of what you have to say in *The Moral Arc*. For example, we largely applaud your definition of "moral progress" as "the improvement in the survival and flourishing of sentient beings." Though we believe that rational beings should take precedence over other sentient beings, nevertheless that is an excellent definition—and one that is very much in keeping with Aristotle, we might add!

You make only one glancing reference to Aristotle in connection with ethics, yet it seems to us that your moral system is clearly a form of eudaimonism—in which morally good or virtuous behavior is grounded in what it means for human beings to flourish as rational animals. If that is right, then aren't you really an Aristotelian at heart? If not, why not?

Michael Shermer: Yes, I'm an Aristotelian, although I graft onto that parts of other moral philosophies, such as natural rights theory and sometimes utilitarianism and occasionally Rawlsian original position theory. No one moral theory can get it right for all circumstances, so we have to cobble together parts of what our greatest minds have generated before us. All I'm trying to do in *The Moral Arc* is establish that: (1) there are objective transcendent moral truths—right and wrong—and these are grounded in nature and human nature; and (2) there is no wall separating *is* and *ought*. Everyone just repeats the naturalistic fallacy without ever reading what Hume actually said—which I did, and then took my interpretation to one of the world's leading Hume scholars, Oxford University philosopher Peter Millican, who confirmed that he thinks my interpretation of Hume is accurate. This is all in Chapter 1 of my book.

James Barham: Although you do not discuss Aristotle, eudaimonism, or virtue ethics in any detail, you do spend a couple of pages discussing the concept of "natural rights" in connection with John Locke as the foundation of your individualist approach to morality, which we applaud. We understand why you wish to claim a direct lineal descent from Locke, in accordance with your claim that Enlightenment "science and reason," not religion, have been the principal drivers of human progress.

Now, natural rights are normally thought of as grounded in natural law—and so ultimately in human nature.[3] Of course, we also understand that the view of human nature underpinning your invocation of natural rights is based on the neo-Darwinian theory of evolution, as you have explained at length in several of your previous books, as well as in *The Moral Arc*. We will explore all of this in detail with you in a few moments. However, first we wanted to point out a significant historical connection that you do not mention. Obviously, Locke himself knew nothing of Darwin. For him, as for the other Enlightenment figures whom you cite, natural rights were principally grounded in the natural law tradition leading back to Hugo Grotius—who lived a couple of generations before Locke—and beyond Grotius to the great sixteenth-century Spanish Scholastics (Francisco de Vitoria, Domingo de Soto, Bartolomé de las Casas, Francisco Suárez, et al.), who in turn based their ethical and political thought on earlier Scholastic philosophy, notably that of Aquinas and Ockham.

In short, the Lockean tradition of human rights which you wish to claim as the offspring of the Enlightenment, we would claim is in reality to a very significant degree the offspring of Scholasticism—i.e., of Christian philosophy. How would you respond?

Michael Shermer: I have taken a number of courses from The Teaching Company on the history of rights and the origin of the concept of natural rights—Rufus Fears's "History of Freedom," Dennis Dalton's "Freedom: The Philosophy of Liberation," and Joseph Koterski's "Natural Law and Human Nature"—all of whom take the concept back to the ancient Greeks. So, you have to start the historical timeline somewhere, or else we'll end up with all ideas as footnotes to Plato, as Whitehead said—wrongly, I might add.

I begin with Locke because that was the most influential source for the founding of America and the modern concept of natural rights as it is understood and practiced today. Certainly, the Scholastics were hugely influential in their time, as were the Renaissance humanists such as Erasmus in their time. But *The Moral Arc* is not a history book meant to convey the full and rich history of ideas, but rather, as you properly discerned, a work with a central thesis in the spirit of what I call Darwin's Dictum: "All observation must be for or against some view if it is to be of any service."

James Barham: Here is another problem we find with your general approach to morality in *The Moral Arc*: You quite rightly point to the importance of what you call "the principle of interchangeable perspectives" as absolutely fundamental to human morality. Your principle appears to mean more or less the same thing as Adam Smith's "impartial spectator," Kant's "categorical imperative," or simply the Golden Rule. That is, in our dealings with others, we ought to give much, though not necessarily overriding consideration to their interests, and not just to our own or those of our family, friends, tribe, etc. With all of this, very few would disagree.

But then you go on to say the following:

Reason and the principle of interchangeable perspectives put morals more on a par with scientific discoveries than cultural conventions. Scientists cannot just assert a claim without backing it up with reasoned argument and empirical data…

But the claim that human morality is closely akin to natural science is problematic, to say the least. For one thing, if it were true, it would suggest that the vast majority of human beings who have ever lived were all egoists blind to the claims of morality—which we take to be an unacceptable consequence of your view. More seriously still, by assimilating reason to science, you seem to be labeling most of humanity as irrational, conflating a highly refined and specialized form of reasoning (natural science) with the general human capacity to reason (common sense). While common sense undoubtedly has its limits, nevertheless it is a thoroughly rational process. Every time a Paleolithic hunter said to himself, "If I want to be successful in the hunt tomorrow, I must sharpen my spear blade," that was human reason in action. And it is this universal commonsense form of reason that, in our view, is at the root of the principle of interchangeable perspectives, not science.

In short, we believe that you are confusing science with reason itself in claiming that morals are "on a par with scientific discoveries." How say you?

Michael Shermer: The Paleolithic hunter who deduces that he must take certain actions to be successful in his hunt is employing a form of scientific reasoning by proposing a hypothesis ("If I want to be successful in the hunt tomorrow, I must sharpen my spear blade") and then testing it the next day to see if it works. That's not a moral matter, but once brains evolved the capacity to substitute parts in an equation ("If I try, X then Y will result, and whenever Y happens, I also did X"), then we can employ that same capacity to reason about other people, our actions and theirs, and the consequences for both. As I wrote in *The Moral Arc*, referencing Steven

Pinker's analysis of the role of reason in moral progress in his book, *The Better Angels of Our Nature*[4]:

> Reason is part of our cognitive makeup, and once it is in place it can be put to use in analyzing problems it did not originally evolve to consider. Pinker calls this an open-ended combinatorial reasoning system that, "even if it evolved for mundane problems like preparing food and securing alliances, you can't keep it from entertaining propositions that are consequences of other propositions." This ability matters for morality because "if the members of species have the power to reason with one another, and enough opportunities to exercise that power, sooner or later they will stumble upon the mutual benefits of nonviolence and other forms of reciprocal consideration, and apply them more and more broadly."

As well, I develop the idea of what I call the "Witch Theory of Causality": If your explanation for why bad things happen is that your neighbor flies around on a broom and cavorts with the devil at night, inflicting people, crops, and cattle with disease, preventing cows from giving milk, beer from fermenting, and butter from churning—and if you believe that the proper way to cure the problem is to burn her at the stake—then you are either insane or you lived in Europe six centuries ago, and you even had biblical support in Exodus 22:18: "Thou shalt not suffer a witch to live."

The witch theory of causality gives us insight into how moral progress is made—by achieving a better understanding of causality. It is evident that most of what we think of as our medieval ancestors' barbaric practices, such as witch burning, were based on mistaken beliefs about how the laws of nature actually operate. If you — and everyone around you — truly believe that witches cause disease, crop failures, sickness, catastrophes, and accidents, then it is not only a rational act to burn witches, it is a moral duty. This is what Voltaire meant when he wrote: "Those who can make you believe absurdities, can make you commit atrocities."

James Barham: As we said before, there is a great deal of *The Moral Arc* that we found not only of absorbing interest, but to be right on target. Your discussion of free will and moral responsibility is a case in point. Here, we find your rehearsal of contemporary brain research to be exemplary, and frankly far more nuanced than most such discussions in either the scientific or the philosophical literature. That said, once again we would beg to disagree that your research supports your general conclusion about the advance of science sustaining the moral arc. In fact, the contrary is arguably the case.

It is of particular interest to us that you develop the concept of "free won't" in some detail. This is the idea that the essence of human freedom is our ability to veto certain impulses that we judge to be morally wrong. But this seems to be little more than the standard bifurcated view of man — "suspended between the beasts and the angels" — familiar from many religious traditions, including Christianity. Religion aside, it is also familiar from the tradition of philosophical reflection upon morality. For example, in the eighteenth century Bishop Butler stressed that human nature is dual in this sense, and that conscience is precisely the capacity of the higher part of our nature to hold the lower part in check.[5]

In short, your account of modern neuroscience merely confirms what faith traditions and common sense have known all along. But far from helping to shore up this traditional view of human moral responsibility, your emphasis on neuroscience is all too liable to be abused in order to undermine freedom and moral responsibility by those of a more reductionist and determinist bent than yourself. And surely anyone who was already having trouble controlling his base urges would be unlikely to be helped by being made to believe that "my neurons made me do it," whereas he might well be helped by understanding his situation in fundamentally religious terms as a heroic struggle between his higher and his lower natures.

We have two questions is this connection: (1) Do you accept that the neuroscience is basically superfluous, merely confirming what we already knew about freedom and moral responsibility?; and (2) Do you agree that your reliance on neuroscience is potentially dangerous in the hands of authors less scrupulous than yourself?

Michael Shermer: My view of human nature is not so different from that of the Christian worldview in which we have a dual nature of good and evil, albeit mine is grounded in a scientific understanding of what those terms mean in the context of human action: Good and evil do not exist outside of human thought and action; there is no "evil" in the world in the theological sense of a force at work separate from humans.

Neuroscience is just a tool to help us understand human thought and action. It can be used by thinkers on either side of the free will/determinism debate, and like any other area of science, it can be misused by people, as in the case of eugenics and Social Darwinism. In such debates, much turns on the definition of terms and how words are used, so to that extent there may be an inherent limitation in our knowledge, which is why I tried to break out of the categorical, either/or mode (either we're free or we're determined), and work toward a more continuous scale of degrees of freedom—healthy humans more than addicted or brain-damaged humans, humans more than monkeys, dogs more than cockroaches, ants more than bacteria. When you take that approach it allows for more subtle and nuanced analysis of particular cases: Charles Manson had less choice in becoming a career criminal than I do, but he's still responsible for his actions because he could have done otherwise nonetheless.

Again, in my moral theory it isn't necessary to explain every last case; only that we can account for most cases most of the time. Our civil society with the rule of law that presumes people make moral choices and we punish them when they choose to break the law, works for most people in most circumstances most of the time. And that law takes into account diminished capacity to choose, as in the case of a crime of passion vs. premeditated murder.

Most of us most of the time in most circumstances are free to choose, and as such we should be held accountable for our actions.

James Barham: Any final thoughts you would like to share with our readers? Where would you like to be 5-10 years from now?

Michael Shermer: In five to 10 years from now I would like to live in a world in which gays and lesbians can marry and have all the same rights as everyone else in this country. I would like to see the continued decline of violence and the expansion of the moral sphere to include even more people as full rights-bearing sentient beings. I would like to see an end to poverty in Africa in 15 years, and the reformation and enlightenment of Islam in 30 years to be fully in accord with Western values that treat women as equals instead of chattel property. I would like to see a colony on Mars, all automobiles gone electric, and everyone on the planet with a device that gives them access to the Internet and all knowledge free and readily available to everyone.

Is that asking too much? Should not a man's reach exceed his grasp, or what's a heaven for?

James Burham: Thank you very much for your time!

PART 1
MATERIALISM IN SCIENCE

Dr. Sheldrake defends the position that science needs to free itself from materialist dogma; indeed, science misunderstands nature by being wedded to purely materialist explanations.

Dr. Shermer opposes Dr. Sheldrake's position, arguing that science, properly conceived, is a materialistic enterprise; for science to look beyond materialist explanations is to betray science and engage in superstition.

Chapter 1

Rupert Sheldrake's Opening Statement

Dear Michael,

We agree about many things. We both think that scientific research and the scientific method are of enormous importance. We both believe in evolution. We share an interest in the history of science. And we are both in favor of skepticism.

Where we differ is in our degree of skepticism. I am more radical than you. I think we need to question the dogmas of science itself. As the physicist Richard Feynman observed, scientists need to find out not only what might be right about their theories, but also what might be wrong with them.

For more than 150 years, scientific orthodoxy has been based on the philosophy of materialism, the claim that all reality is material or physical. All of our own experiences are by-products of physical and chemical activities in our brains. Even God exists only as an idea in human minds, and hence in human heads. Brains are made up of unconscious matter and governed only by impersonal physical and chemical laws. Like all other features of living organisms, they have evolved through chance mutations and natural selection, without any purpose or direction.

These beliefs are powerful not because most scientists think about them critically, but because they don't. The facts of science are real enough, and so are the techniques that scientists use, and so are the technologies based on them. But the beliefs that govern conventional scientific thinking are an act of faith, grounded in a 19th-century ideology.

Contemporary scientific orthodoxy rests on the following assumptions:

1. **Everything is essentially mechanical.** Dogs, for example, are complex mechanisms, rather than living organisms with goals of their own. Even people are machines, "lumbering robots" in Richard Dawkins's vivid phrase, with brains that are like genetically programmed computers.

2. **All matter is unconscious.** It has no inner life or subjectivity or point of view. Even human consciousness is an illusion produced by the physical activities of brains.

3. **The total amount of matter and energy is always the same** with the exception of the Big Bang, when all of the matter and energy of the universe suddenly appeared.

4. **The laws of nature are fixed.** They are the same today as they were at the beginning, and they will stay the same forever.

5. **Nature is purposeless**, and evolution has no goal or direction.

6. **All biological inheritance is material**, carried in the genetic material, DNA, and in other material structures.

7. **Memories are stored as material traces in brains** and are wiped out at death.
8. **Minds are inside heads and are nothing but the activities of brains.** When you look at a tree, the image of the tree you are seeing is not "out there," where it seems to be, but inside your brain.
9. **Unexplained phenomena like telepathy are illusory.**
10. **Mechanistic medicine is the only kind that really works.** Many people are unaware that these doctrines are assumptions; they think of them as science, or simply believe that they are true. They absorb them by a kind of intellectual osmosis.

In my book, *Science Set Free: 10 Paths to New Discovery* (Deepak Chopra, 2012), I tried the experiment of turning these assumptions into questions, treating them as testable scientific hypotheses rather than dogmas. None stood up very well to the evidence. And all led to further questions—some of which I would like to ask you, Michael. My questions are *in italics*.

1. Is nature mechanical?

The mechanistic theory of nature gives a supreme privilege to machine metaphors. The genes are programs; the heart is a pump; the brain is a computer. But many aspects of nature are not machine-like, including the entire universe. The theory that everything started in a Big Bang resembles ancient myths of the hatching of the cosmic egg. Ever since it hatched, the universe has been growing

47

and developing ever more form and diversity; it seems much more like a developing organism than a machine. And developing plants and animals themselves, like oak trees or cats, are more like true organisms, with their own goals and purposes, than purposeless machines.

Michael, do you think of yourself as a complex machine or as a conscious living organism?

2. Is matter unconscious?

The Scientific Revolution of the 17th century gave birth to modern science by creating a radical dualism between unconscious matter and conscious, non-material minds possessed only by humans, angels, and God. This duality was mirrored in the more or less peaceful coexistence of religion, the arts, and the sciences from the 17th century to the 19th century. Religion and the arts were concerned with conscious experience, while the realm of science was the physical universe.

In the 18th and 19th centuries, many people reacted against the power of churches by becoming atheists, especially in countries like France and Russia, where the established churches were allied with reactionary, authoritarian governments. Materialism provided arguments in support of atheism, and gave it scientific credibility. By denying the existence of immaterial consciousness, atheistic materialists got rid of God and angels at one stroke. There were no longer two realms of reality, matter and consciousness; there was only one reality, matter.

Materialism became the predominant orthodoxy of science by the end of the 19th century.

Materialists got rid of God, or at least confined him to the brains of believers, but they were left with the "hard problem" of explaining how unconscious matter becomes conscious within human brains. This problem continues to haunt the neurosciences and the philosophy of mind. If consciousness is an illusion, or nothing but a byproduct of brain activity, it cannot actually do anything, and hence we cannot make free choices.

Michael, do you believe that you have free will?

3. Is the total amount of matter and energy always the same?

The constancy of the total amount of matter and energy made sense in the eternal universe of 19th-century physics, when eternal laws of nature governed an eternal physical reality. Most materialists still believe in changeless laws of nature and constant amounts of matter and energy, with one exception: All of the matter and energy in the universe sprang from nothing at the moment of the Big Bang. Leaving aside this miraculous origin, the nature of matter and energy is not straightforward. Physicists now postulate that about 96 percent of reality is made up of dark matter and dark energy, whose nature is literally obscure.

Michael, do you believe that the total amount of dark matter and dark energy is always the same (except at the moment of the Big Bang)?

4. Are the laws of nature fixed?

The idea of fixed "laws of nature" is a hangover from pre-evolutionary cosmology, which prevailed until the Big Bang theory became orthodox in 1966. In an evolving universe the laws themselves may evolve, or they may be more like habits—as I myself think, as the American philosopher C.S. Peirce suggested in the early 20th century, and as the contemporary cosmologist Lee Smolin also proposes.

Michael, if you believe that all the laws and constants of nature came into being fully formed at the moment of the Big Bang, how does the universe remember them? Where are they imprinted?

5. Is nature purposeless?

The assumption that nature is purposeless follows from the machine metaphor. Machines have no purposes of their own, but only those imposed upon them or programmed into them to serve human purposes. If the universe is mechanical, then evolution is purposeless: It has no goals, intentions, or direction.

Michael, is the purposelessness of evolution a testable hypothesis?

6. Is all biological inheritance material?

Genes have been greatly overrated. They do not "code for" or "program" the form and behavior of organisms, like the shape of an orchid flower or the nest-building instincts of a weaverbird. They specify the sequence of amino acids in protein molecules, and some genes are involved in the control of the activity of other genes.

The Human Genome Project has been disappointing because it was based on a false conception of what genes do. The "missing heritability problem" is now provoking a crisis in modern biology because it turns out that as much as 70 percent of inheritance does not appear to be explained by genes. Also, the recognition of the epigenetic inheritance at the beginning of this century means that the inheritance of acquired characteristics, once taboo, is now mainstream. For example, recent experiments have shown that mice can inherit the fears of their fathers. Male mice were made averse to the smell of a synthetic chemical, acetophenone, by being given electric shocks while they smelled it. Their sperm were used to fertilize female mice, and their children and grandchildren were terrified of the smell of acetophenone.

Findings of this kind mean we need to modify Neo-Darwinian evolutionary theory, which was based on the primacy of genes and a denial of the inheritance of acquired characteristics. Neo-Darwinism assigned creativity to random mutations of genes. But if organisms can learn and adapt to their environments, and pass on adaptations to their offspring, then evolution is affected by organisms' own abilities to learn and adapt.

Michael, how have your views of evolution changed in the light of epigenetic inheritance?

7. Are memories stored as material traces?

The assumption that memories are stored as material traces in brains has dominated scientific research for more than a century, but the hypothetical memory traces have proved surprisingly elusive. There is plenty of evidence that particular parts of the brain become active when memories are being laid down and retrieved, but where they exist in between is mysterious. As I suggest in my book *Science Set Free*, memories may depend on a kind of resonance across time. Brains may be more like TV receivers, tuning into memories transmitted from their own past, than like video-recorders. Brain damage can affect the retrieval of memories, just as damage to a TV set can affect the sounds or pictures it produces, but this does not prove the damage has destroyed a storage system.

There is also a philosophical problem: The theory that memories are stored in material traces means there must be a retrieval system that recognizes the memories it is trying to retrieve. To recognize the memories, the retrieval system must itself have a memory. And if it has a material memory, then the retrieval system itself needs a retrieval system, and so on.

Michael, doesn't this standard explanation of memory either presuppose memory, or fall into an infinite regress?

8. Are minds confined to brains?

If minds are nothing but the activity of brains, then they must be confined to the insides of heads. But when I look at a tree, I do not experience the image of the tree inside my head; I experience it where the tree is. My image of the tree is in my mind, but it is not inside my head. Our minds may be extended beyond our brains every time we look at something.

Everyone agrees that vision involves light coming into eyes, causing activity in the retina, impulses up the optic nerves, and specific patterns of activity in brains. Most materialists assume that the nervous tissue then somehow generates a 3-D, full-color virtual reality display inside the skull. But since these virtual reality displays are invisible to objective observers, how do we know they are inside skulls? Instead, we may generate images that are projected out to where they seem to be. Our minds may reach out to touch what we are seeing— for example, a tree. We may affect what we are looking at. That may be why many people and animals often sense when they are being watched, even when looked at from behind. And, in my opinion, there is good evidence for the reality of the sense of being stared at, which we will probably discuss in our next dialogue.

Meanwhile, Michael, how do you interpret your own experience of seeing? When you look at the sky, do you think that you are seeing the sky inside your skull?

9. Are unexplained phenomena like telepathy illusory?
We will be discussing telepathy and other psychic phenomena in our next dialogue.

10. Is mechanistic medicine the only kind that really works?
I think the best way of evaluating different kinds of therapy is through comparative effectiveness research, pragmatically finding out which therapeutic systems work best for common problems like lower back pain or migraine headaches. Some patients, selected at random, would be sent to regular physicians, others to acupuncturists, chiropractors, osteopaths, homeopaths, and practitioners of other systems that claim to offer cures. Which systems work best? And which are most cost-effective? Through pragmatic research, we can have an evidence-based approach to medicine without a commitment to one particular system or ideology.

Michael, do you think that governments and health insurance companies should fund research comparing the effectiveness of different kinds of therapy, including alternative therapies?

Like you, Michael, I am pro-science. But we have different ways of expressing our enthusiasm. I think the interests of the sciences are best served by exploring what we do not understand, even if that leads us beyond the limitations imposed by the materialist philosophy. My scientific allegiance is not to a particular worldview, materialism, but to science as a method of inquiry, open to new possibilities.

Some atheists fear that if they let go of materialism they will allow back religion, and set off a tsunami of superstition. Therefore, materialism must be defended at all costs. But this fear is exaggerated. Atheists need not be shackled to materialism; they can move beyond it and still remain atheists. For example, the distinguished American philosopher Thomas Nagel rejects materialism in favor of panpsychism in his book, *Mind and Cosmos: Why the Materialist, Neo-Darwinian Conception of Nature Is Almost Certainly False* (Oxford University Press, 2012). He proposes that there are aspects of mind throughout the natural world, even in the chemical elements. He also argues that evolution is purposive. But he remains an atheist.

Whether God exists or not is a different question, which we will discuss in our third dialogue. Our present discussion is about whether the sciences ought to remain within the materialist paradigm, or whether scientists should be free to range more widely.

—*Rupert*

* * *

Michael Shermer Responds

Dear Rupert,

Scientists in general, and myself in particular, are not materialists because of historical contingencies related to French and Russian reactionary governments, or because we're afraid that if we give up materialism, religion and belief in God will come roaring back. Last time I checked the polls, in fact, religion and belief in God seemed to be doing just fine in this age of scientific materialism, although I am encouraged by the recent increase in the number of people with no religious affiliation. But my desire to see the power of religion attenuated has nothing to do with scientific materialism and everything to do with human rights and moral progress (which I claim is primarily driven by science and reason and Enlightenment humanism). And I am an atheist not because I am a materialist, but because I do not believe there is sufficient evidence for the existence of God.

Materialism became the predominant worldview of science by the end of the 19th century because it works—it enables scientists to search for and find mechanistic explanations for a wide spread of phenomena, from atoms and molecules to ecologies and economies. This is not an act of *faith*, as you say, but of *confidence* built over centuries of data-gathering, hypothesis-testing, and theory-building, all contested through the competitive enterprise of science in which skeptics have, as you note in quoting Feynman, tried to find out what might be wrong with their own and especially others' ideas.

Neither is materialism an ideology, as you also suggest. An ideology is a set of beliefs about how society should be structured, and, traditionally, those beliefs have had less to do with science and more to do with preconceived notions of the proper place of people in a society (usually held by the dominant group as a way of keeping minority groups in their place—or eliminating them altogether). Communism and National Socialism, for example, were not scientific societies (as often claimed by theists in their eagerness to indict atheism by linking it to science and materialism), but utopian societies grounded in the faux religions of Marxism (for Communist states) and nationalism (for the Nazis). Also, in the case of the Nazis, their eugenics program had nothing to do with scientific materialism or atheism, and their science was (as I write in *The Moral Arc* [p. 137]) "a thin patina covering a deep layer of counter-Enlightenment, pastoral, paradisiacal fantasies of racial ideology grounded in ethnicity and geography."

As seen in these examples, much of our dialogue turns on the meaning of words. My example from my first letter about the discovery of the material mechanism of the nerve impulse illustrates the point. And just as we no longer have to depend on fuzzy phrases like "nervous energy" or *force mentale* (because we now understand the underlying molecular process), we should be cautious when we employ fuzzy words like "thought" or "mind" so as not to reify them into causal explanations. It's okay to say "I think" or "my mind" in conversation, as long as we all understand that these are just linguistic place-fillers for a complex electro-chemical exchange going on inside our brains.

Such linguistic clarification goes a long way to answering your many questions. "Is nature mechanical?" As opposed to what—non-mechanical? What would that mean? What would be an example of a non-mechanical system? Yes, scientists use metaphors like the heart as a pump or the brain as a computer, but metaphors are just a way of talking and thinking about something with the end goal of understanding the underlying mechanical processes. Communication is easier when I say "my heart pumps blood" or "my brain computes the consequences," instead of providing a long description of muscle contractions and nerve impulses.

You ask whether I see myself as a complex machine or as a conscious living organism? Yes. Both. But, again, these are just words, and you have used them in a manner that implies one can't be both. A conscious living organism is a complex machine—very complex. As well, being a "complex machine" in no way detracts from the elegance and beauty of being a "living organism"; it only adds to it. This reminds me of the 19th-century English poet John Keats, who once bemoaned that Isaac Newton had "destroyed the poetry of the rainbow by reducing it to a prism." Natural philosophy, he lamented in his 1820 poem "Lamia,"

... will clip an Angel's wings
Conquer all mysteries by rule and line
Empty the haunted air, and gnomed mine—
Unweave a rainbow ...

Richard Feynman gave as good a response as anyone to Keats in his book, *What Do YOU Care What Other People*

Think? Further Adventures of a Curious Character (W.W. Norton, 1988), in his contemplation of a flower:

> The beauty that is there for you is also available for me, too. But I see a deeper beauty that isn't so readily available to others. I can see the complicated interactions of the flower. The color of the flower is red. Does the fact that the plant has color mean that it evolved to attract insects? This adds a further question. Can insects see color? Do they have an aesthetic sense? And so on. I don't see how studying a flower ever detracts from its beauty. It only adds.

The sense I get from your work—and that of many others who harbor some reservations about science under the rubric of "scientism"—is that the reductionistic, mechanistic worldview somehow detracts from both the beauty and understanding of nature. It doesn't, it only adds by giving us a deeper understanding of it.

You ask if matter is unconscious. The answer depends on the meaning of the word "conscious." If you mean something akin to what you and I are doing here—consciously thinking and experiencing and communicating—then what matter are we talking about? Is an apple (the fruit, not the computer) conscious? If you mean to ask if apples can think, experience, and communicate, then obviously not. But I wouldn't even say that apples are unconscious because that implies that they have a temporary suspension of consciousness, like when we are put under anesthesia. Consciousness is not part of an apple's essence, so it can't even be unconscious, and neither can the molecules with which it is made.

You ask if I have free will. I do. I have an entire chapter in *The Moral Arc* explaining why, but in brief I present four ways around the paradox of retaining freedom and moral responsibility in a determined universe:

- *Modular mind*—even though a brain consists of many neural networks in which one network may make a choice that another network finds out about later, they are all still operating in a single brain
- *Free "won't"*—vetoing competing impulses and choosing one thought or action over another
- *Degrees of moral freedom*—a range of choice options varying by degrees of complexity and the number of intervening variables
- *Choice as part of the causal net*—wherein our volitional acts are part of the determined universe but are still our choices.

But the free-will issue also turns on how we define these terms. I'm guessing that you too believe in free will, but probably for different reasons involving something akin to a conscious or spiritual or nonmaterial soul or entity or substance that represents "you" that is making choices. But this doesn't give you free will. It just means something else is making the decision for "you," unless you think that this other entity is you and the physical entity called Rupert Sheldrake is something else.

You ask about matter, energy, and the laws of nature. I'm not a physicist and am not qualified or even sufficiently read to offer a proper response. My friend and colleague Lawrence Krauss, a highly respected cosmologist and physicist, tells me (in an email dated May 1), "We know that over cosmic time the fine structure constant has been constant by at least 1 part in 100,000 or so, and the gravitational constant hasn't changed by more than 40 percent at most since the universe was one second old. The latter comes from BBN [Big Bang Nucleosynthesis] limits, and the former from measuring the spectrum of light emitted by atoms in galaxies at high redshift." Lawrence's book, *A Universe from Nothing: Why There Is Something Rather than Nothing* (Atria Books, 2013), along with Stephen Hawking and Leonard Mlodinow's book, *The Grand Design* (Random House Publishing Group, 2010), presents materialist/mechanistic models by which a universe can come into existence and sustain itself without the necessity of a higher intelligence or outside creative force of any kind.

As for "laws of nature," again, linguistic precision is helpful here. "Laws" are the linguistic and mathematical descriptions we humans give to naturally occurring repeating phenomena. There are no laws of nature "out there." Nature just *is*. Stars, for example, convert hydrogen into helium in a well-defined manner dependent on temperature and pressure. We can write out the mathematical equations that tell us how this happens, how fast, how much, and so on. But there are no "laws" inside stars; just material stuff doing what it must do under those conditions.

And this answers your next question, "Is nature purposeless?" What do you mean by purpose? If you mean that the purpose of stars is to convert hydrogen into helium under certain temperatures and pressures, then yes, nature has purpose. Stars are fulfilling their "destiny" in this sense. But if you mean by "purpose" some outside transcendent source that grants or directs purpose, or that acknowledges or rewards purpose fulfilled, then no such source exists. There is no Archimedean point from which we can lever into our lives some external purpose. We have to create our own purpose, and we do this by fulfilling our nature, by living according to our essence, by being true to ourselves, as Shakespeare wrote:

> This above all: to thine own self be true,
> And it must follow, as the night the day,
> Thou canst not then be false to any man.

—*Michael*

* * *

Rupert Sheldrake Replies

Dear Michael,

Thank you for your response to the questions I asked in my Opening Statement.

At first I was puzzled by the contradictions in your positions. But then I read your chapter on free will in *The Moral Arc* (Henry Holt, 2015). I was struck by your endorsement of the theory of modular minds, according to which we have many different compartmentalized brain functions, or mental apps (as on a smart phone). As you put it, "There is no unified 'self' that generates internally consistent and seamlessly coherent beliefs devoid of conflict. ... Instead, we are a collection of distinct but interacting modules that are often at odds with one another" (p. 338).

Suddenly, your inconsistencies became easier to understand. There are several different Shermer modules that predominate in different contexts:

- *The libertarian module*, with a passionate belief in the importance of freedom and individual autonomy. This gives you a strong incentive to argue for the reality of freedom and individual autonomy. "Morality involves conscious choice," as you make clear in *The Moral Arc* (p. 335), which presupposes free will. This belief in freedom also inspires you to quote scientists like Richard Feynman and Robert Oppenheimer, who advocate free inquiry. Indeed, *The Moral Arc* opens with Oppenheimer's words: "There must be no barriers to freedom of inquiry. There is no place for dogma in science."

I agree. But other Shermer modules pull your brain in different directions:

- *The scientific materialist module*, powered by a faith in scientific authority. You dismiss problems that cast doubt on this worldview as verbal quibbles, easily solved by reasserting materialist beliefs. For example, as you wrote in your Response, thought and mind "are just linguistic place-fillers for a complex electro-magnetic exchange going on inside our brains." But if everything is determined by physical causes, and consciousness is nothing but an aspect of physical brain activity, materialism denies the existence of any genuine freedom of choice, a conclusion accepted by many of your fellow atheists and materialists, including Sam Harris. However, a denial of free will is in conflict with your libertarian and secular humanist modules. So, in *The Moral Arc*, the lack of free will in a materialist universe is dissolved away, as if by verbal magic, in seven pages. Free will is reinstated to your own satisfaction, enabling the libertarian and materialist modules to coexist in your brain, even though they seem at odds with each other.

- *The philosophy of science module.* As you say, according to some philosophers of science, laws of nature are not "out there"; they are just human-made descriptions and mathematical models. But this view conflicts with the views of your friend, fellow atheist Lawrence Krauss, and a highly respected cosmologist. He speculates that the universe arose out of nothing in accordance with the laws of quantum physics, and presupposes that laws were already there at the beginning of the universe, long before there were any humans to describe them. This is surely a metaphysical assumption. Moreover, in an evolutionary universe, if the laws are just descriptions of regularities, the laws must evolve because the regularities of nature evolve.

- *The secular humanist module.* As the atheist philosopher John Gray puts it in his book *Straw*

Dogs (Granta Books, 2003), "Humanism is not science but religion—the post-Christian faith that humans can make a world better than any in which they have so far lived. Humanism is the transformation of the Christian doctrine of salvation into a project of universal human emancipation." Gray argues that nothing in atheism or materialism or neo-Darwinism supports this optimistic faith. If the implications of these theories are taken seriously, "we are animals like any other; our fate and that of the rest of life on Earth are the same." The optimistic humanist module is very different from the materialist module, even though they coexist in your brain, and in millions of others.

- *The crusader module.* Crusaders were warriors, fighting against infidels and heretics. Organized skeptical movements are modern secular crusades, defending scientific orthodoxy, opposing heresy, and expanding the influence of scientific materialism. But crusaders and their successors, the inquisitors, were not noted for encouraging free inquiry, especially if it was directed towards the foundations of their faith. The crusader module is inherently intolerant.

Luckily, when your libertarian module is uppermost, we agree that scientific inquiry should be free, and not constrained by dogma. We will return to this subject next month.

—Rupert

Chapter 2

Michael Shermer's Opening Statement

Dear Rupert,

We have never met in person, but I have been following your work ever since your book *The Presence of the Past: Morphic Resonance and the Habits of Nature* (Times Books, 1988) was published. I was in a doctoral program in the history of science at the time, and I recall being intrigued with your thesis that the past's influence on the present goes beyond the usual socio-cultural effects that historians track (such as political, economic, and cultural forces that carry across the centuries— the *longue durée*, as the French *Annales* school of historians calls it). Clearly, you meant something more than the strict materialist forces at work that scientists study (including historical scientists, under which heading I include historians along with archaeologists), and this is where my skeptical alarm went off as I tried to understand what mechanism within the known laws of nature could possibly be at work for the present to be influenced by the past in ways that you suggest.

As you know and have been critical of—most recently in your book *The Science Delusion* (Coronet, 2012; titled *Science Set Free* in the United States)—most scientists (myself included) adopt the materialist position of methodological naturalism, which I take to mean: Life is the result of a natural and purposeless process in a system of material causes and effects that does not allow, or need, supernatural forces. In my public talks I often illustrate the principle with the famous Sidney Harris cartoon of two scientists at a blackboard filled with equations in which the words "THEN A MIRACLE OCCURS" appear in the middle of the mathematical sequence. The caption has one scientist saying to the other: "I THINK YOU NEED TO BE MORE EXPLICIT HERE IN STEP TWO."

This is sometimes called the "God of the gaps" argument—wherever an apparent gap exists in scientific knowledge, this is where we interject a miracle from God as an explanation. It works something like this, when dealing with certain biological features of organisms, in which "X" may be the eye or DNA or some other feature:

1. X looks designed.
2. I can't think of how X was designed naturally.
3. Therefore, X was designed supernaturally.

This fallacy reminds me of the "plane problem" of Isaac Newton's time: The planets all lie approximately in a plane (known as the ecliptic). Newton found this arrangement to be so improbable that he invoked God as an explanation in *Principia Mathematica*: "This most beautiful system of the sun, planets, and comets could only proceed from the counsel and dominion of an intelligent and powerful Being." Since Newton's time, however, that gap has been filled in with natural explanations of how stars and solar systems are formed from condensing clouds of interstellar gas in which eddies of material conglomerate into a central star (or two, in the case of binary stars) and multiple smaller planets (Kant was the first to propose this "nebular hypothesis"). Creationists do not cite this problem or quote Newton because the gap has now been filled in with a natural explanation by scientists.

The materialism of methodological naturalism also bothers the Intelligent Design movement because of their desire to introduce supernaturalism into the system of the world. For example, in his book, *Darwin on Trial* (Regnery Gateway Publishing Company, 1991), the University of California-Berkeley law professor Phillip E. Johnson—one of the founders of the Intelligent Design movement—accused scientists of unfairly defining God out of the picture by limiting the search to only natural causes. He charged that scientists who postulate that there are non-natural or supernatural forces or interventions at work in the natural world are being pushed out of the scientific arena on the basis of nothing more than a fundamental rule of the game. Like you, Johnson and his Intelligent Design colleagues such as William Dembski, Paul Nelson, and Stephen Meyer want the rules of the game changed to allow methodological supernaturalism.

Let's play out that scenario and imagine what methodological supernaturalism would look like in science, and how it would work. (I did this in my book *Why Darwin Matters* [Times Books, 2006].) For the sake of argument, let's assume that Intelligent Design theorists have discovered a new force of nature that accounts for the apparent design in such features as the eye or DNA. How will they identify it? Will it be considered a new natural force, or a new supernatural force? By what criteria will they discriminate between the natural and the supernatural? How can one tell?

For example, in the early 20th century, the British biologist Julian Huxley parodied the French philosopher Henry Bergson's fuzzy explanation for life as being caused by an *élan vital* (vital force), which Huxley said was like explaining a railroad steam engine as being driven by its *élan locomotif* (locomotive force). In his book, *The Ancestor's Tale* (Houghton Mifflin Harcourt, 2004), Richard Dawkins employed a similar analogy to parody Intelligent Design explanations for life. To say that the eye or DNA are "designed" tells us nothing. Scientists want to know how they were designed, what forces were at work, how the process of development unfolded, etc. Dawkins imagined a counterfactual history in which Andrew Huxley and Alan Hodgkin, winners of the Nobel prize for figuring out the molecular biophysics of the nerve impulse, in a creationist frame of mind, attribute it instead to "nervous energy."

Along these same lines (inspired by Dawkins's analogy, which I first employed in my book *The Believing Brain* [Times Books, 2011]), imagine if David Hubel and Torsten Wiesel—winners of the 1959 Nobel Prize for their pioneering research in brain circuitry and determining the neurochemistry of vision—had, instead of spending years getting down to the cellular and molecular level of understanding how the brain converts photons of light into neural impulses, simply attributed the process to a *force mentale*.

Now see here, Hubel, this business about how photons of light are transduced into neural activity is a dreadfully thorny problem. I just can't understand how it works, can you?

No, my dear Wiesel, I can't, and implanting those electrodes into monkeys' brains is truly unpleasant and messy, and I have the hardest time getting the electrode into the right spot. Why don't we just say that the light is converted into a nerve impulse by a *force mentale*?

What does invoking a concept like *force mentale* explain? Nothing. It would be like describing your automobile's engine as operating by a "combustive force," which fails to capture what is actually going on inside the cylinders of an internal combustion engine: A piston compresses a vaporous mixture of gasoline and air that is ignited by a spark plug causing an explosion that drives the piston down thereby turning a crank arm that is connected to a drive shaft that is linked to a differential that rotates the wheels. Giving something a label like "nervous energy," *force mentale,* or "combustive force," is not an explanation. It is just a label to talk about something material that is at work that we want to try to understand with natural forces.

In this sense, then—as I've mentioned many times in my critique of theories about psi, ESP, miracles, and the like—there is no such thing as the paranormal or the supernatural. These words, "paranormal" and "supernatural," are precisely parallel to "nervous energy" and *force mentale*: just linguistic placeholders to talk about something for which we do not as yet have a normal or natural explanation.

Analogously, when cosmologists talk about "dark energy" and "dark matter," they don't mean those words to be an explanation, only linguistic placeholders until they figure out what exactly is causing such as-yet unsolved mysteries such as the rotation of galaxies and the accelerating expansion of the cosmos. Whereas cosmologists do not stop searching for the underlying mechanisms of the observed phenomena just because they have a label, paranormalists and supernaturalists treat words like "paranormal" and "supernatural" as if they were causal explanations. They're not.

Turning to your area of research, Rupert, if it turned out that, say, people really could read other people's minds and that they were able to do so because (*pace* Roger Penrose's and Stuart Hameroff's theory) inside our neurons are tiny microtubules in which quantum effects happen that allow thoughts (patterns of neural firing) to be transferred from one skull to another at any distance (like the "spooky action at a distance" effects that quantum physicists have measured in experiments), that would not be ESP or PSI, and we would not need to call it a "paranormal" effect, because we would then know that the ability to read minds was due to the properties of neurons and atoms. If this turned out to be true (I'm skeptical), this new theory would be subsumed under the sciences of neuroscience and/or quantum physics (quantum neuroscience?) and would no longer be studied under the umbrella of, say, parapsychology.

This is not a new problem. Scientists and philosophers of science have long struggled with defining what constitutes legitimate scientific knowledge, and no less a mind than the great British astronomer Sir Arthur Stanley Eddington chimed in on the debate in his 1939 classic work *The Philosophy of Physical Science* (recall that it was Eddington who successfully tested Einstein's theory of relativity by measuring the bending of starlight by the sun during a solar eclipse in 1919). Eddington made this analogy that I use in my critical thinking course (and in my book *Why People Believe Weird Things* [MJF Books, 1997]), which I think has some relevance to our discussion on the nature of science:

> Let us suppose that an ichthyologist is exploring the life of the ocean. He casts a net into the water and brings up a fishy assortment. Surveying his catch, he proceeds in the usual manner of a scientist to systematize what it reveals. He arrives at two generalizations:
> 1. No sea-creature is less than two inches long.
> 2. All sea-creatures have gills.
> ... In applying this analogy, the catch stands for the body of knowledge which constitutes physical science, and the net for the sensory and intellectual equipment which we use in obtaining it. The casting of the net corresponds to observation ...
> An onlooker may object that the first generalization is wrong. "There are plenty of sea-creatures under two inches long, only your net is not adapted to catch them." The ichthyologist dismisses this objection contemptuously. "Anything uncatchable by my net is *ipso facto* outside the scope of ichthyological knowledge." In short, "what my net can't catch isn't fish."

Extending the analogy beyond the physical sciences to all fields, regardless of what forces may be at work in our world, if they can be measured by our scientific instruments (or by our senses), then by definition they must be natural forces (regardless of what you call them). In other words, what our senses and scientific nets catch are natural fish.

Later, we will be discussing God, but in this context, let me note that if one were to argue that God exists outside of our world (or outside of the universe, or outside of nature), and that God's forces are non-natural (or supernatural) and they can still affect the world but in a non-measurable way (because our scientific nets only catch natural fish), then what's the difference between an invisible God and a nonexistent God?

And if God (or some creative force—it need not be the creator Judeo-Christian-Muslim God) exists outside of nature, but periodically reaches into our world to change it in some manner (such as answering prayers or performing miracles), then, in principle, there should be some way to measure such effects (e.g., patients who are prayed for heal faster, or a physically impossible feat occurs, such as the regrowth of an amputated human limb) and deduce that the source of the effects is outside of all known natural forces. In that case, in principle, such a God (or force) would simply become part of the natural world (at least when He/She/It operates on it).

Thus, it seems to me that once we have carefully defined our terms, it is clear that there really is only the material world, methodological naturalism is the only means to understand it, and science is the only form of reliable knowledge that we have.

—*Michael*

* * *

Rupert Sheldrake Responds

Dear Michael,

I agree with you that science is not about the supernatural. If things can be investigated by the natural sciences, they are part of nature. I support the principle of methodological naturalism, and in my own research have always worked within it.

In relation to psychic phenomena, like you, I have long argued that if they occur (which I think they do), they are natural, not supernatural; normal not paranormal. I also consider morphic resonance— which I suggest underlies memory in nature—to be normal, not paranormal; natural, not supernatural.

Also, like you, I'm against the concept of intelligent design, but for different reasons. I think the word "design" has misleading mechanistic implications. The old version of intelligent design was that God was outside of nature, and designed the machinery of the world, like an engineer designing a machine, or a watchmaker designing a watch. To say that living organisms are "designed" implies that they are complex machines. New versions of intelligent design are subtler, but still imply that living organisms are machines, and that their complexity is designed by a supernatural mind or minds outside nature. I agree with advocates of intelligent design in thinking that evolutionary creativity goes beyond blind chance, but I see living organisms as organisms, not machines, and I think that creativity is inherent in nature, rather than being imposed upon it from outside.

I also agree with you in rejecting "God of the gaps" arguments for the existence of God. We will return to a discussion of God in our third dialogue. But, unlike you, I am skeptical of "materialism of the gaps" arguments. Instead of invoking God, many materialists try to solve problems by making scientific promises. For example:

- "We do not yet fully understand how genes program the development of animals and plants, but in the light of the spectacular advances of molecular biology and vastly improved computer modelling techniques, we soon will."
- "We do not yet know the detailed mechanisms whereby the brain is programmed to produce consciousness, but with the spectacular advances in brain imaging techniques, rapid advances in our understanding are imminent."

Empty phrases such as "genetic programs" and "brain mechanisms" are used to explain almost everything. They imply that the answers are known in principle, leaving only the details to be worked out, with the answers only a few years (or decades) away. Committed materialists are committed precisely because they believe that materialistic explanations will be found in the future. They put their trust in what they hope for—in what is not yet known. The philosopher of science Karl Popper called this attitude "promissory materialism," because it involves issuing undated promissory notes for future discoveries. Promissory materialism is a faith.

In the 19th century, materialism seemed quite straightforward. Old-style materialists thought that matter was made up of hard, enduring stuff, with atoms like little billiard balls, pushed around by known forms of energy. But the nature of physical reality—which materialists think of as the only reality—is much more problematic today. Quantum theory has dissolved matter into vibratory patterns of activity within fields. And most cosmologists and astronomers believe that about 96 percent of the universe is made up of dark matter and dark energy, whose nature is literally obscure. These names may be placeholders, but what they mean is that 96 percent of what materialists or physicalists believe in is unknown. How can we be sure that dark matter and dark energy— the basis for the existence of galaxies and the evolution of the universe—are completely mindless and unconscious?

The most interesting contemporary debate within the materialist community is between conservative materialists—among whom I think you, Michael, are numbered—and animistic materialists, who propose that there are mind-like properties throughout the natural world, even in electrons. As you know, this position is usually called "panpsychism" (from the Greek words pan ["all" or "everything"] and psychē ["soul"]— meaning "all or everything is soul"). For example, the philosopher Galen Strawson argues that materialism itself implies panpsychism. He is a panpsychist, but still thinks of himself as a materialist or physicalist. The neuroscientist Christof Koch has recently come to the conclusion that a version of panpsychism modified for the 21st century is "the single most elegant and parsimonious explanation for the universe." The philosopher Thomas

Nagel, in his book *Mind and Cosmos* (Oxford University Press, 2012), is another eminent proponent of the idea that there is psyche or mind throughout nature. None of these panpsychists proposes that psyches or minds are supernatural, but rather that they are aspects of nature. I agree with them.

I myself think that part of the mental aspect of nature is memory. My own particular hypothesis is that memory depends on the process I call morphic resonance, an influence of similar patterns of activity on subsequent similar patterns of activity, resonating through or across space and time. This resonance occurs in self-organizing systems, which include molecules, crystals, plants, animals, social groups, planets, solar systems, and galaxies. Similar self-organizing patterns of activity resonate across time, from the past to the present. Each species has a kind of collective memory. Every individual both taps into this collective memory and contributes to it. By contrast, human-made objects like chairs, or cars, computers, and other machines are not self-organizing. They are designed and made by people, or through computer programs designed by human computer programmers. They are not subject to morphic resonance from past chairs, cars, etc., although the molecules and crystals within them are.

Morphic resonance is a hypothesis, not an accepted scientific fact. But the progress of science depends on exploring hypotheses, and testing them empirically. The hypothesis of morphic resonance leads to many testable predictions. For example, if rats learn a new trick in New York, rats in London and Sydney should learn the same trick quicker, even in the absence of any conventional means of communication. There is already evidence that this effect occurs. There are many other lines of evidence that seem to support this hypothesis, as summarized in the third edition of my book *A New Science of Life* (Icon Books, 2009; retitled *Morphic Resonance* in the United States [Park Street Press, Rochester VT, 2009]). I have also suggested several new tests in the realms of low temperature physics, crystallography, developmental biology, animal behavior, and human psychology. In biology, this hypothesis implies that the inheritance of form and behavior depends largely on morphic resonance, rather than on genes, which code for the sequence of amino acids in proteins. Genes are not "programs" for development or for instincts. Indeed, it turns out that about 70 percent of human heritability is not explicable in terms of genes. This is called the "missing heritability problem."

One of the most striking implications of morphic resonance concerns memory. Morphic resonance depends on similarity. The greater the similarity, the stronger the resonance. Think about yourself. Which organism in the past was most similar to you? Surely you yourself! Self-resonance is the most powerful resonance working on any self-organizing system. In living organisms, self-resonance helps maintain their form, even though the chemicals and cells within them are continually turned over and replaced.

In the realms of learning and mental activity, self-resonance underlies memory. In other words, memories may not be stored in brains. Brains may be more like TV receivers than video recorders. TV receivers tune in to invisible resonances across space, transmitted through invisible radio waves. Your memories may depend on a resonance with yourself in the past, transmitted across time by morphic resonance. The standard assumption is, of course, that memories are stored as material traces inside brains. But after 100 years of intensive research, these traces have proved extraordinarily elusive, perhaps because they are not there. If I came to your house and analyzed the wires and transistors of your TV set to try and find out what you were watching last week, I would be disappointed; I would find no material traces.

But doesn't the fact that brain damage can lead to loss of memory prove that memories are stored in brains? No. It only shows that properly functioning brains are necessary for the retrieval of memories. Damage to a TV set can lead to changes in the sounds or the pictures, but this does not prove that what you are seeing and hearing is stored inside the set.

I suppose that, in the end, most of our disagreements about science come down to our different agendas. I am a research scientist, and I like exploring new possibilities. You are a leader of the organized skeptic movement, many of whose members are conservative materialists, dedicated to maintaining materialist law and order, patrolling the frontiers of science, and ringing alarm bells. These differences divide us. But what may bring us closer is a belief in free inquiry. I was interested to read in your TheBestSchools. org interview about your libertarian sympathies, which presumably include a belief in the freedom of the sciences from authoritarianism and dogma.

—*Rupert*

* * *

Michael Shermer Replies

Dear Rupert,

In response to your second letter, and your point about seeing living organisms as "organisms, not machines," again, I ask: Why can't they be both? An organism is a living machine, and a complex enough machine can become a living organism! Proponents of Artificial Intelligence (AI) would likely agree for future machines that reach a certain level of intelligence—the singularity, say—at which point these machines will be indistinguishable from living organisms in their actions and cognitions, even though we could lift the hood and see that they are just complex machines. So we're really talking about *degrees of complexity* here, and at some point a machine can become so complex that it appears for all intents and purposes to be alive and organic, as our intuitions understand those concepts. (And future AI will force us to revise those intuitions, along with our legal and moral systems about what constitutes a sentient being deserving of rights and personhood.)

Now, we both agree that these complex organic machines were not designed from the top down (or from the outside by an intelligent designer), so the problem to explain is the source of the complexity, the creative spark behind the complex design. Stephen Hawking once famously asked, "What is it that breathes fire into the equations?" And he equally famously answered "no one"—the universe comes equipped with fire already built into the equations. You and I seem to agree that there is no "who" on the outside breathing life into inorganic matter, so the answer to the question about the source of life's fire must come from within.

You argue that "creativity is inherent in nature." I agree, if by creativity you mean that certain laws of nature—in particular those laws governing biology, embryology, epigenetics, genetics, etc.—lead organisms to unfold embryologically from a tiny cluster of cells into a full-fledged organism (e.g., a mammal), and for species to evolve from simple to complex (and even for the evolution of evolvability). That is, inherent in what Aristotle called the "final cause" of a thing (in this case, an organism) is what it is, by nature, destined to be—a seed to become a plant, a human embryo a human being—under the right conditions. Just as a star is destined to convert hydrogen into helium under the right conditions of heat and pressure, a seed or an embryo must become a plant or human under the right conditions that allow the processes of genetics, epigenetics, embryological development, and the like to unfold as they are destined to do by the laws governing the actions of their molecules.

As you know, the great German philosopher and writer Goethe developed a biological theory of morphology (he invented the word!) based on the formalist idea that the wide diversity of plant and animal complexity can be reduced to single archetypes or forms—a "leaflike" form for leaves, for example. But Goethe's formalist morphological theory never panned out because no underlying mechanism to drive it was ever found, much less the source of the archetype in the first place. Darwin's genius was to turn this theory on its head by showing how all current forms are derived from prior forms, modified by natural selection to be adaptive to current environments. Thus, the "archetype" of our arms and hands—the tetrapod forelimb with a humerus, ulna, and radius, and carpals, metacarpals, and phalanges bones—evolved in fish 375 million years ago as an adaptation for transitioning from the water onto the land (amphibians), most famously in the *Tiktaalik* fossil discovered by Neil Shubin in 2004. And fossils before and after that type ("transitional fossils") show no "archetype" at all, only local adaptations to local environments all the way back and forward from that particular type.

So we already have a purely materialist explanation for the diversity of types in nature without the need to invoke the "memory" of forms (in your theory) that guide molecules toward them (if I'm understanding your theory correctly), but if you are right in your claim that "genetic programs" do not explain forms, body types, anatomy, or physiology, then I have three questions:

1. What do you think DNA is for and what is it doing, if not what geneticists think it is doing?
2. Where is the "memory" for forms (say, the tetrapod forelimb) stored?
3. How does this memory act on physical systems, such as the molecules that make up cells or organs or (in keeping with my example) tetrapod forelimbs?

I think the reason your theory of morphic resonance has not gained acceptance within the scientific community is the same as why Goethe's formalist theory of morphology never succeeded: In science, we need both theory and mechanism. Alfred Wegener had a theory of drifting continents but no mechanism that could drive plates around the globe. Once that mechanism was found in the 1960s in the form of plate tectonics, the theory gained acceptance. It's true that Darwin didn't have an understanding of genetics as a mechanism for natural selection, but the evidence for his theory was so overwhelming from so many different lines of inquiry that it gained acceptance despite this shortcoming. And, of course, after 1953, genetics synthesized the theory of evolution into the fully mature science it is today.

For your theory to follow a similar trajectory, you would need to:

1. Provide more reliable and consistent (i.e., replicable) evidence that such memory in nature exists;
2. Explain where this memory is stored (i.e., what's the mechanism of storage); and
3. Explain how morphic memories affect physical systems (e.g., molecules).

If you were able to do these things, then I and most everyone else in science would change our minds and accept your theory.

Until then, it is reasonable to be skeptical.

—*Michael*

PART 2
MENTAL ACTION AT A DISTANCE

Dr. Shermer argues that psychic or psi phenomena are artifacts of poor experimental procedure or outright fraud; no convincing evidence or experiments support their reality.

Dr. Sheldrake defends the position that telepathy, ESP, and psychic/psi phenomena in general are real and backed up by convincing evidence; their investigation deserves to be part of science.

Chapter 3

Michael Shermer's Opening Statement

Dear Rupert,

As a précis to our second subject of "Mental Action at a Distance," allow me to respond to several points in your third letter as they set the stage for what is to come.

First, throughout our dialogue you have mentioned several times my "libertarian" perspective, which you presume to mean that I should be open minded to ideas outside the mainstream of science, such as your own. In the words of Inigo Montoya (in the film *The Princess Bride*), "You keep using that word. I do not think it means what you think it means." On this side of the pond, libertarian is a political position affiliated with individual rights and small government; libertarians tend to be socially liberal and fiscally conservative. Libertarianism has nothing to do with open mindedness as a feature of human cognition, nor with the "libertarian" position in the free will-determinism debate.

Second, you reference many people in contradistinction to my positions. On the free will/determinist issue, for example, you note that my friend Sam Harris is a determinist. So what? My friend Dan Dennett is a compatibilist (as is my friend Steve Pinker). The argument from authority in either direction is fallacious.

I've carved out my own position on free will in *The Moral Arc*: that one can be a scientific materialist and still believe in human volition, which you failed to refute.

Third, you cite my friend Lawrence Krauss as a refutation to my claim that "laws of nature" are just descriptions of regularities in nature. Once again, this is not an argument against my position, nor did you defend your own (presumably counter) position. In like manner you cite John Gray's definition of humanism as a "religion," and from this you assert that atheism, materialism, and neo-Darwinism are in conflict with this "faith." Who cares how John Gray defines humanism? He doesn't speak for humanists, most of whom are atheists, materialists, and neo-Darwinians, and in any case, this is not a counter-argument to my position, just an argument from authority. What counts are arguments not authorities. For example, I presented arguments and challenges to your theory of morphic resonance, which I hope you will address, most notably:

1. Where is the "memory" for morphic forms stored (e.g., the tetrapod forelimb)?
2. How does this memory act on physical systems, such as the molecules that make up cells or organs or (in keeping with my example) tetrapod forelimbs?
3. Provide reliable and consistent (i.e., replicable) evidence that such memory in nature exists.

As well, skeptics are not "crusaders ... fighting against infidels and heretics." We're critical thinkers applying science and reason to any and all claims. You, for example, are a skeptic of the materialist-determinist-reductionist paradigm in science, but that doesn't make mainstream scientists infidels and heretics! Organized skeptical movements that have spontaneously emerged around the world are interested in understanding and explaining phenomena on the borderlands of science (e.g., ESP), primarily because most scientists are too busy working in their own fields to devote the necessary time to properly analyze these claims. We're not closed-minded so much as conservative (cognitively, not politically) in offering our provisional assent that a claim is factually true. The reason for this cautiousness is that most claims people make are not true. The history of science is littered with failed hypothesis. For every Galileo whose ideas were borne out by the data and whose theories changed the world, there are thousands of scholars and scientists whose conjectures and speculations failed to generate any supportive evidence.

Like most scientists, we skeptics assume the *null hypothesis* that a claim under investigation is not true (null) until proven otherwise, and the burden of proof is on you to provide convincing experimental data to reject the null hypothesis. And this brings me to the topic of our second set of exchanges about mental action at a distance. Take ESP and a simple example I employed in my book *The Believing Brain: From Ghosts to Gods to Politics and Conspiracies—How We Construct Beliefs and Reinforce Them as Truths* (Times Books, 2011): determining through extra-sensory means whether a playing card randomly selected from a deck is red or black. The null hypothesis is that it is not possible to do this and thus to reject the null hypothesis we would need to establish a figure for the number of correct hits. By chance, we would expect a test subject to get about half correct. In a deck of 52 cards, half of which are red and half of which are black, random guessing or flipping a coin will produce, on average, 26 correct hits. Of course, as gamblers know, there are streaks and deviations from perfect symmetry. The roll of a roulette wheel will not produce a perfect red-black-red-black sequence. Typically, streaks of red and black turn up, often more of one than the other in any given limited sequence, without any violation of the laws of probability.

So for a proper test, we need to run multiple trials in which some rounds may result in slightly below chance (e.g., 22, 23, 24, or 25 hits) and other rounds may result in slightly above chance (e.g., 27, 28, 29, or 30 hits). The variation may be even greater and still due to chance. What we need to determine is the number by which we can confidently reject the null hypothesis. In this example, that number is 35. The subject would need to get 35 correct hits out of a 52-card deck in order for us to reject the null hypothesis at the 99 percent confidence level. Even though 35 out of 52 doesn't sound like it would be that hard to get, by chance alone it would be so unusual that we could confidently state ("at the 99 percent confidence level") that something else besides chance was going on here.

What might that something else be? It could be ESP. But it could be something else as well. Perhaps our controls were not tight enough. Maybe the subject was getting the card color information by some other normal sensory (as opposed to extra sensory) means of which we're not aware. I've seen magicians do something very similar to this test with a deck of cards in which they do far better than 35 out of 52. They get 52 out of 52. What is the probability of that? It is 100 percent because it's a magic trick! That I do not know how the trick is done does not mean that ESP was at work. It just means that we need to be very careful in our controls to insure that we are measuring what we think we are measuring. This is especially true when we're attempting to measure effects far more subtle and complex than the color of playing cards, such as your many claims related to morphic resonance: phantom limbs, homing pigeons, crossword puzzles, how dogs know when their owners are coming home, and how people know when someone is staring at them. Each of these is a separate effect that may or may not have the same set of causes. Consider the claim that people have a sense of being stared at. First, we have to control for the well-known reverse self-fulfilling effect: a person suspects being stared at and turns to check; such head movement catches the eyes of would-be starers who then turn to look at the staree, who thereby confirms the feeling of being stared at. But this is a normal sensory phenomenon, not an extra-sensory phenomenon.

In 2000, John Colwell from Middlesex University, London, conducted a formal test utilizing your suggested experimental protocol, with 12 volunteers who participated in 12 sequences of 20 stare or no-stare trials each, with accuracy feedback provided for the final nine sessions. The results were that the subjects were able to detect being stared at only when accuracy feedback was provided, which Colwell attributed to the subjects learning what was, in fact, a nonrandom presentation of the experimental trials.

And as you well know, when Richard Wiseman attempted to replicate your research, he found that subjects detected stares at rates no better than chance. This led him to believe that for those experiments that did generate statistically significant results there was an experimenter bias problem, which he demonstrated in a collaborative study with Marilyn Schlitz, who is a believer in ESP (Wiseman is a skeptic). They found that when Schlitz did the staring she found statistically significant results, whereas when Wiseman did the staring he found chance results.[1] I found a similar bias effect in a content analysis I did of the 2005 special issue of the *Journal of Consciousness Studies* devoted to your research,[2] in which I rated the 14 open-peer commentaries on your target article (on the sense of being stared at) on a scale of 1 to 5 (critical, mildly critical, neutral, mildly supportive, supportive). Without exception, the 1s, 2s and 3s were all traditional scientists from mainstream institutions, whereas the 4s and 5s were all affiliated with fringe and pro-paranormal institutions. Of course, you might reasonably argue that it is Wiseman and these mainstream scientists whose skeptical bias prevents the effect from being measured (and not *vice versa*), but in this case, since it is you making the extraordinary claim, the burden of proof is on you to provide extraordinary evidence that it is experimenter bias preventing the effect, which in my opinion has yet to be produced.

Pulling back for a historical perspective, ever since organizations such as the Society for Psychical Research were founded in the late 19th century, thousands of experiments have been run in an attempt to measure ESP and related phenomena. Although I know you disagree, most scientists remain unconvinced by the handful of positive findings, noting that the vast majority of experiments failed to reject the null hypothesis that ESP does not exist. Even experiments that did produce statistically significant effects (meaning that they rejected the null hypothesis) were often fraught with methodological shortcomings. Richard Wiseman and Julie Milton, for example, tested your hypothesis in a study at the University of Hertfordshire called "Can Animals Detect When Their Owners Are Returning Home?" To your credit, you made this test possible after a dog owner (and her dog, Jaytee) were featured on a television show as successful examples confirming your hypothesis (that you also published in a paper). But as Wiseman and Milton concluded after instituting tighter controls: "Analysis of the data did not support the hypothesis that Jaytee could psychically detect when his owner was returning home."

Consider one of the most thorough reviews of this literature ever conducted by the highly respected social scientists Daryl Bem and Charles Honorton, entitled "Does Psi Exist? Replicable Evidence for an Anomalous Process of Information Transfer," published in the prestigious review journal *Psychological Bulletin* in 1994.[3] The scientists conducted a meta-analysis, a statistical technique that combines the results from many studies to look for an overall effect, even if the results from the individual studies were nonsignificant (i.e., they were unable to reject the null hypothesis at the 95 percent confidence level). Bem and Honorton concluded: "The replication rates and effect sizes achieved by one particular experimental method, the *ganzfeld* procedure, are now sufficient to warrant bringing this body of data to the attention of the wider psychological community."

The ganzfeld procedure places the "receiver" in a sensory isolation room with ping pong-ball halves covering the eyes, headphones playing white noise over the ears, and the "sender" in another room attempting to transmit photographic or video images via ESP (or psi). Despite finding evidence for psi—subjects had a hit rate of 35 percent when 25 percent was expected by chance—Bem and Honorton lamented: "Most academic psychologists do not yet accept the existence of psi, anomalous processes of information or energy transfer (such as telepathy or other forms of extrasensory perception) that are currently unexplained in terms of known physical or biological mechanisms" (p. 118). Why don't scientists accept psi despite this apparent significant effect? I contend that there are two reasons: *data* and *theory*.

Data. Both the meta-analysis and *ganzfeld* techniques have been challenged by scientists. Ray Hyman from the University of Oregon found inconsistencies in the experimental procedures used in different ganzfeld experiments (that were lumped together in Bem's meta-analysis as if they used the same procedures), and that the statistical test employed (*Stouffer's Z*) was inappropriate for such a diverse data set. Hyman also found flaws in the target randomization process (the sequence in which the visual targets were sent to the receiver), resulting in a target selection bias: "All of the significant hitting was done on the second or later appearance of a target. If we examined the guesses against just the first occurrences of targets, the result is consistent with chance."[4] Julie Milton and Richard Wiseman conducted a meta-analysis of 30 more *ganzfeld* experiments and found no evidence for psi, concluding that psi data are non-replicable, a fatal flaw in scientific research.[6] In general, over the course of a century of research on psi, the tighter the controls on the experimental conditions, the weaker the psi effects seem to become, until they disappear entirely. This is a very strong indicator that ESP is not real.

Theory. The deeper reason scientists remain skeptical of psi—and will even if more significant data are published—is that there is no explanatory theory for how psi works. Until psi proponents can explain how thoughts generated by neurons in the sender's brain can pass through the skull and into the brain of the receiver, skepticism is the appropriate response. If the data show that there is such a phenomenon as psi that needs explaining (and I am not convinced that they do), then we still need a causal mechanism.

Consider Bem's more famous 2011 study of psi entitled "Feeling the Future: Experimental Evidence for Anomalous Retroactive Influences on Cognition and Affect," published in the prestigious *Journal of Personality and Social Psychology*. Bem sat subjects in front of a computer screen that displayed two curtains, behind one of which would appear a photograph that was neutral (e.g., a building), negative (e.g., a car accident), or erotic (e.g., sex). Through 36 trials, the subjects were to *preselect* which screen they thought the image would appear behind, *after* which the computer randomly chose which window to project the image. When the images were neutral, the subjects did no better than 50/50 guessing. But when the images that were about to be projected were erotic in nature, the subjects preselected the correct screen 53.1 percent of the time, which Bem reports as statistically significant. Bem calls this "retroactive influence"—erotic images ripple back from the future into the present, or as the comedian Stephen Colbert called it when he featured Bem on his show *The Colbert Report*, "Extra Sensory *Porn*ception." When Colbert pressed Bem for an explanation for how such time reversal could possibly work, the scientist confessed, "We have no idea," but then suggested quantum physics as a possible mechanism. Once again, there are many reasons to be skeptical—six to be precise.

1. The journal also published in the same issue a paper by the psychologist Eric-Jan Wagenmakers that was critical of Bem's findings, concluding that his methodology was flawed in that he was using an exploratory analysis of psi hypotheses to see what might turn out significant, and then presenting it as if it were confirming the hypothesis. (See Wagenmakers' publication list on his Web page.)

2. Bem's study is an example of what I call negative evidence: If science cannot determine the causes of X through normal means, then X must be the result of paranormal causes. Ray Hyman, who has devoted his career to carefully analyzing serious psi research, calls this issue the "patchwork quilt problem" in which "anything can count as psi, but nothing can count against it." That is, "If you can show that there is a significant effect and you can't find any normal means to explain it, then you can claim psi." That is not a valid conclusion, however, especially when dealing with such extraordinary claims as ESP.

3. Paranormal effects, which are rarely detected at all, are always so subtle and fleeting as to be useless for anything practical, such as predicting the future, locating missing persons, gambling, investing, and so on.

4. A small but *consistent* effect might be significant (for example, in gambling or investing), but according to Ray Hyman, Bem's 3 percent above-chance effect in Experiment 1 was not consistent across his nine experiments, which measured different effects under varying conditions.

5. Experimental inconsistencies plague such research. For example, Hyman notes that in Bem's Experiment 1 the first 40 subjects were exposed to equal numbers of erotic, neutral, and negative pictures, but then Bem changed the experiment in midstream and for the remaining subjects just compared erotic pictures with an unspecified mix of all types of pictures. Plus, it turns out that Bem's fifth experiment was conducted before his first, which raises the possibility that there might be a post-hoc bias in either running the experiments or in reporting the results. As well, Bem reports that "most of the pictures" were selected from a source called the *International Affective Picture System*, but he doesn't tell us which ones were not, why, or what procedure he employed to classify images as erotic, neutral, or negative. Hyman's list of flaws numbers in the dozens. As he told me in an interview for one of my *Scientific American* columns

(May, 2011): "I've been a peer reviewer for over 50 years, and I can't think of another reviewer who would have let this paper through peer review. They were irresponsible."

6. Perhaps they missed what York University psychologist James Alcock found in another paper that Bem wrote, entitled "Writing the Empirical Journal Article" (posted on his website), in which Bem instructs students: "Think of your data set as a jewel. Your task is to cut and polish it, to select the facets to highlight, and to craft the best setting for it. Many experienced authors write the results section first."

In other words, Bem began with what he presumed to be true and then worked backward to find the data to fit it. This is called the confirmation bias, and it has plagued psi research for over a century. Thus it is I remain skeptical.

—*Michael*

* * *

Rupert Sheldrake Responds

Dear Michael,

When it comes to science, you are not only conservative, as you admit, but also authoritarian. If the science establishment is for it, you are for it. If the science establishment is against it, you are against it.

But when it comes to the U.S. political system, as a libertarian you want less government and more individual freedom and autonomy. The reason I referred to your libertarian stance was because I hoped that your belief in freedom might also apply to science. After all, much institutional science is a branch of big government. The U.S. government spends about $135 billion a year on scientific research and development. Most scientists are working within highly bureaucratic systems. Yet the advance of science depends on the freedom of inquiry. At present, this freedom is inhibited by institutional orthodoxies. That is why I think we need more individual freedom within science, and less authoritarianism.

You end your statement by saying that "confirmation bias ... has plagued psi research for over a century." But, ironically, your own opening statement is itself a perfect example of confirmation bias. The authorities you quote are all materialists and committed skeptics. In particular, when discussing the data from psi research, you rely on the claims of James Alcock, Richard Wiseman, and Ray Hyman, all of whom are skeptical crusaders and members of the Committee for Skeptical Inquiry. Such skeptical advocacy organizations often behave as if they are running election campaigns, with all the vices of confrontational party politics: bias, *ad hominem* attacks on opponents, negative campaigning, and attempts to conceal unwelcome facts by muddying the waters. (Many examples are highlighted on the website SkepticalAboutSkeptics.org.)

The same techniques are widely used by denier movements and campaigning skeptics in other fields of activity. For example, product defense lawyers are often hired by corporations to undermine the scientific basis for government regulations. One of the pioneers of this strategy was the cigarette company Brown and Williamson, which ran a campaign to discredit evidence about the harmful effects of smoking, which helped hold back anti-smoking regulations for years. As one of the company's executives commented, "Doubt is our product since it is the best means of competing with the body of fact." David Michaels, who was then the assistant secretary for environment safety and health at the U.S. Department of Energy, pointed out in his article "Doubt Is Their Product," published in the June 1, 2005 issue of *Scientific American* that the same strategy has been adopted by numerous industries making toxins. When confronted with evidence that their products are lethal, the offending industry hires skeptics to muddy the waters. As Michaels noted, "Their conclusions are almost always the same: The evidence is ambiguous, so regulatory action is unwarranted."

When we come to the actual evidence for psi phenomena, you adopt a classic muddying-the-water strategy. To take specific examples you raised in your opening statement:

1. You make it sound as if psi researchers are so naïve they do not know about null hypotheses. In fact, almost all psi research is based on this principle. You yourself are naïve about this or, worse, intentionally misinforming our readers here. If you were to read some actual research papers, instead of reiterating claims by fellow skeptics, you would see that your claims are not true. For example, you could look at null hypothesis testing in my own papers on the sense of being stared

at, telepathy, and the unexplained powers of animals (which I discuss on my website).

2. The skeptic Ray Hyman has, as you say, repeatedly criticized psi research using the ganzfeld technique. However, in 1986, psi researchers worked together with Hyman to produce a joint communiqué on improved procedures that could eliminate possible flaws. In response to these new guidelines, parapsychologists carried out new, computer-controlled versions of the ganzfeld experiment, called the *autoganzfeld*. Skeptics tried to find new flaws, and once again the procedures were tightened up. By 2011, there had been 59 studies in 15 different laboratories following the rigorous methods agreed upon with skeptics. The overall hit rate was very significantly above chance.[1] It is true that in 1999 Richard Wiseman and Julie Milton published a study claiming that the combined results of new ganzfeld studies were not significantly above chance. They reached this conclusion by omitting some recent highly successful experiments. When these were included, the combined results were indeed significantly above chance, as Milton later admitted.[2]

You write, "In general, over the course of a century of research on psi, the tighter the controls on the experimental conditions, the weaker the psi effects seem to become, until they disappear entirely." This vague, evidence-free generalization is simply not true. It is wishful thinking.

3. Daryl Bem's study of "feeling the future" has indeed been attacked by skeptics and again you try to muddy the

waters by bringing up quibbles over statistical details. And while some skeptics have not been able to replicate some of Bem's results, other researchers have. Whether or not presentiment exists is an intriguing open question, and there are several other lines of investigation besides Bem's that strongly suggest it does.[3]

Incidentally, if you want to attack Bem for advising students to write their results section first, you should attack most other scientists as well, because this is standard procedure. For example, the current guidelines to PhD students at Cambridge University state: "Write your chapters in the following order: Results, Methods, Discussion, Introduction."

4. You attempt to portray the skeptical claims of Richard Wiseman and Julie Milton as a refutation of a dog's ability to know when his owner was coming home, when in fact their data supported it. You made no mention of my refutation of their claims, nor to independent analyses by several investigators, all of whom concluded that Wiseman and Milton's claims were highly misleading.[4]

5. Instead of looking at the actual data from experiments on the sense of being stared at, you ranked the status of the institutions of the scientists who commented on it, and found that those from the most mainstream institutions were most critical. This is surely no more than an argument from authority, and has nothing to do with actual evidence. In the joint experiments carried out on staring by Marilyn Schlitz and Richard Wiseman, as you rightly say, Wiseman found only chance results when he did the staring, whereas when Schlitz did the

staring there were statistically significant results. But the experimenter effects in this experiment were not symmetrical. The procedure involved staring through closed circuit television in Wiseman's own laboratory, under conditions that eliminated any possible sensory information. Schlitz's positive results could not have occurred as a result of wishful thinking or bias. On the other hand, Wiseman could easily have obtained no effect by not staring very hard, and indeed he later said he had found it "an enormously boring experience" and that in most of the trials he was "pretty passive about it."[5]

But, as you admit, this debate is not really about evidence. Committed skeptics are against psi phenomena because they do not fit in with the materialist worldview. This is dogmatic, not scientific. There are already several hypotheses as to how psi may work, but they offend your authoritarian instincts because they go beyond existing scientific orthodoxy. My own hypothesis of morphic fields, for example, may help to explain telepathy. I do not have space here to answer your questions about how I think morphic fields and morphic resonance may work, but a summary can be found on my website.

When Michael Faraday first proposed his hypothesis of electric and magnetic fields, he could not explain how they worked. It was another 20 years before James Clerk Maxwell came up with a theory in terms of the ether, a form of "subtle matter," which was itself unexplained. And then, in 1905, Albert Einstein showed that the ether did not exist and came up with yet another theory. Fortunately, organized skeptic groups did not exist at the time of Faraday. If they had done, his research would have been dismissed on the grounds that

MENTAL ACTION AT A DISTANCE

his invisible "fields" could not be explained in terms of existing mechanistic theories, and Maxwell would have been treated as a pseudo-scientist for suggesting the existence of invisible "subtle matter."

Reactionary skepticism does not advance the cause of science. It inhibits scientific inquiry, and shuts down curiosity. There is already a great deal of evidence that psi phenomena exist, despite the tireless efforts of crusading skeptics to misdirect attention and pretend that psi effects have "disappeared entirely."

In the end, our differences come back to our different roles. You are a professional skeptic trying to uphold the authority of science, guard its frontiers, and root out heresy. I am a research scientist trying to explore unexplained phenomena.

—*Rupert*

* * *

Michael Shermer Replies

Dear Rupert,

In your last letter you accuse me of being a "committed skeptic" who is "against psi phenomena because they do not fit in with the materialist worldview." You keep repeating this point despite my protests to the contrary. As I've said all along, if psi were real and we understood how it worked, then it would be part of the scientific worldview explainable by natural forces, even if that means expanding our understanding of what constitutes "natural." Again, there is no supernatural; just natural and mysteries that remain unexplained by natural causes. It is possible that one day psi will be accepted by the scientific community and incorporated into the scientific worldview (even the materialistic worldview), but at the moment it isn't, for both evidentiary and causal reasons.

I have reviewed the many shortcomings with your claimed evidence. In perusing your website to understand your causal theory involving morphic resonance, I find a number of flaws in your conjectures, starting with the name itself. "Morphic resonance" is too broad a category—too generalized—and it attempts to explain too many separate phenomena (that very likely have separate causes) to be a useful theory. How can a single field of any sort explain cellular membranes and microtubules; the body types of dogs (from Afghan hounds to poodles); memory in rat; social groups (from schools of fish to flocks of birds); human customs such as the Jewish Passover, the Christian Holy Communion, and crossword puzzles; perceptual phenomena such as vision (the sense of being stared at); and even the laws of nature and the Big Bang origin of the universe?! Is there anything this theory can't explain? I contend that any theory purporting to explain everything in effect explains nothing.

In your last letter, you cite Michael Faraday, James Clerk Maxwell, and their theory of fields explaining electric and magnetic phenomena, as analogous to your own morphic resonance fields. First, Faraday and Maxwell were attempting to explain one specific phenomenon that involved electricity and magnetism and their relationship; they were not attempting to explain everything from inheritance to cultural customs to human memory to the origin of the universe! And it wouldn't have mattered if there was an organized skeptical movement or not, because in time their theory proved correct irrespective of what any skeptical evaluator might have thought about it. Likewise, the skepticism you have encountered has not primarily been from organized skeptical groups, but rather from mainstream working scientists who have no affiliation with organized skepticism. In any case, skepticism is inherent to the scientific process itself, as I noted previously in arguing that all scientists start with the skeptical position of the null hypothesis that their experiments attempt to reject. The burden of proof is on you.

You make the analogy with fields "extending beyond the material objects in which they are rooted" — such as magnets and magnetic fields, planets and gravitational fields, and cell phones and "cellular" electromagnetic fields—to argue that the mind extends beyond the brain. The flaw in this reasoning is that these other fields are detectable, measurable, quantifiable, and predictable. What detectable evidence do you have of the mind extending beyond the brain? The equivalent of a Geiger counter that detects the radiation extending beyond radioactive materials would go a long way toward supporting your morphic resonance theory of mind. To date, you appear to have no such detectable evidence of such a field, and instead rely on subjective feelings people have about being stared at or receiving phone calls or emails from people whom they are thinking about, which have other equally plausible explanations (more plausible in my thinking).

As well, surely the morphic resonance field that directs the development of cellular microtubules is different from the field that shapes the DNA-protein chain sequence, and different still from the field that controls memory, cultural artifacts, and the evolution of species. You do not seem to have any evidence for any such fields, outside of a "god-of-the-gaps" type argument that if a scientist can't explain phenomenon X (e.g., embryological development), then a morphic field must be the explanation. It's not enough to challenge the prevailing theory of X; you must proffer your own testable explanation for X and provide evidence for it. In my opinion, you have yet to do so, and so I and most scientists remain skeptical.

Finally, it appears to me that your morphic resonance theory is circular: "Morphic fields contain other morphic fields within them"; morphic fields "contain a built-in memory given by self-resonance with a morphic unit's own past and by morphic resonance with all previous similar systems"; and so forth. It seems to me you are explaining morphic resonance fields with …morphic resonance fields. It would be tautological to assert that morphic fields cause morphic fields. What is the cause of morphic resonance fields in the first place?

—*Michael*

Chapter 4

Rupert Sheldrake's Opening Statement

Dear Michael,

For committed materialists, psychic (psi) phenomena such as telepathy and the sense of being stared at must be illusory because they are impossible. Minds are inside brains. Mental activity is nothing but electro-chemical brain activity. Hence thoughts and intentions cannot have direct effects at a distance, nor can minds be open to influences from the future. Although psi phenomena *seem* to occur, they must have normal explanations in terms of coincidence, subtle sensory cues, wishful thinking, or fraud.

Dogmatic skeptics often repeat the slogan that "extraordinary claims demand extraordinary evidence." But the sense of being stared at and telepathy are not extraordinary, they are ordinary. Most people have experienced them. From this point of view, the *skeptics'* claim is extraordinary. Where is the extraordinary evidence that most people are deluded about their own experience? Skeptics can only fall back on generic arguments about the fallibility of human judgment.

I here consider research on the sense of being stared at and telepathy. These are subjects on which I have published more than 20 peer-reviewed research papers in scientific journals, as well as two books, *Dogs That Know When Their Owners Are Coming Home* (2nd ed.: Three Rivers Press, 2011) and *The Sense of Being Stared At: And Other Unexplained Powers of Human Minds* (2nd ed.: Park Street Press, 2013). Because of limitations of space, I omit the other three main areas of psi research: clairvoyance or remote viewing, meaning the ability to see or experience things at a distance; precognition or presentiment, knowing or feeling a future event; and psychokinesis, or mind-over-matter effects.

The detection of stares

Most people have felt someone looking at them from behind, turned around, and met the person's eyes. Most people have also experienced the converse: They have sometimes made people turn around by staring at them. In extensive surveys in Europe and North America, between 70 percent and 97 percent of adults and children reported experiences of these kinds. Many species of non-human animals also seem able to detect looks. Some hunters and wildlife photographers are convinced that animals can detect their gaze even when they are hidden and looking at animals through telescopic lenses or sights.

If the sense of being stared at is real, then it must have been subject to evolution by natural selection. How might it have evolved? The most obvious possibility is in the context of predator-prey relations. Prey animals that can detect when predators are looking at them will stand a better chance of surviving than those that cannot.

Since the 1980s, the sense of being stared at has been investigated experimentally both through direct looking and also through closed-circuit television. In the scientific literature it is variously referred to as "unseen gaze detection" or "remote attention" or "scopesthesia" (from the Greek *skopein*, meaning "to view," and *aisthēsis*, meaning "perception"). The majority of these studies, even some of the studies by skeptics, have shown positive, statistically significant effects. The largest experiment on the sense of being stared at began in 1995 at the NEMO Science Centre in Amsterdam. More than 36,000 people took part, with positive results that were astronomically significant statistically. The most sensitive subjects were children under the age of nine.

Telepathy in real life

In most, if not all, traditional societies, telepathy seems to be taken for granted and put to practical use. For example, many travelers in Africa reported that people seemed to know when people to whom they were attached were coming home, even though they had no normal means of knowing. The same occurred in rural Norway, where there is a special word for the anticipation of arrivals: *vardøger*. Typically, someone at home heard a person approaching the house, and coming in, yet nobody physically did so. Soon afterwards the person really arrived. Similarly, the "second sight" of the some of the inhabitants of the Scottish Highlands included visions of arrivals before the person actually arrived.

In an attempt to find out more about telepathy in modern societies, I launched a series of appeals for information through the media in Europe, North

America, and Australia. Over 20 years, I have built up a database of seemingly telepathic and other psi experiences containing more than 10,000 case histories. Many people have observed responses of animals like dogs and cats to their thoughts and intentions. The most impressive occur when the people are miles away, far beyond the range of the animals' senses. More than 1,500 people have told me that their dogs and cats know when a member of the family is coming home and go to wait at a door or window, often 20 minutes or more in advance. Many of these stories make it clear that the animals' responses were not simply reactions to the sounds of a familiar car or familiar footsteps in the street. They happened too long in advance, and they also happened when people came home by bus or train. Nor was it just a matter of routine. Some people, like plumbers, lawyers, and taxi drivers, worked irregular hours, and yet the people at home knew when they were coming because the dog or cat went to wait at a door or window.

Among humans, many cases of apparent telepathy occur in response to other people's needs. For example, hundreds of mothers have told me that when they were breastfeeding, they knew when their baby needed them, even from miles away. They felt their milk let down. The milk letdown reflex is mediated by the hormone oxytocin, sometimes called the love hormone, and is normally triggered by hearing the baby cry. The nipples start leaking milk and many women feel a tingling sensation in their breasts. When nursing mothers were away from their babies and felt their milk let down, most of them took it for granted that their baby needed them, even though it was not at a routine feeding time. They were usually right. They did not experience their milk letting down because they started thinking about the baby; they started thinking about their baby because their milk let down for no apparent reason. A statistical analysis showed that this was not a matter of synchronized physiological rhythms.

A telepathic connection between mothers and their babies makes good sense in evolutionary terms. Mothers who could tell at a distance when their babies needed them would tend to have babies that survived better than babies of insensitive mothers.

Until the invention of modern telecommunications, telepathy was the only way in which people could be in touch at a distance instantly. In most respects, telepathy has now been superseded by telephones—but it has not gone away. Indeed, telepathy now occurs most commonly in connection with telephone calls.

Experimental research on telepathy with animals

I started my own research on telepathy with animals, rather than people. I thought that if telepathy occurs, it is normal, not paranormal, natural, not supernatural, and is likely to have evolved as a communication system between bonded members of animal groups.

In particular, I carried out many experiments with return-anticipating dogs to find out whether they really did anticipate their owners' returns when they could not have known by "normal" means. In these tests, the owners of the dogs went at least five miles away from home. While they were out, the place where their dog waited was filmed continuously on time-coded videotape. The owner did not know in advance when she would be going home and she did so only when she received a message from me via a telephone pager at a randomly selected time. She traveled by taxi or in another unfamiliar vehicle to avoid familiar car sounds. The dog I have investigated most, a terrier called Jaytee, was at the window on average only 4 percent of the time during the main period of his owner's absence, and 55 percent of the time when she was on the way back. This difference was very significant statistically. The Science Unit of Austrian State Television, ORF, filmed this independent experiment and you can find it on my website.

At my invitation, a leading British media skeptic, Richard Wiseman, did his own videotaped tests with Jaytee under the same conditions. His data showed that Jaytee was at the window 4 percent of the time during the main period of his owner's absence, and 78 percent of the time when she was on the way back, a statistically significant positive effect. However, in the media, Wiseman misleadingly tried to portray this replication of my results as a refutation of Jaytee's abilities!

Experimental research on human telepathy

Since 1880, there have been hundreds of studies of human telepathy under laboratory conditions. The great majority have given positive, statistically significant results.

The commonest kind of apparent telepathy in the modern world concerns telephone calls. Surveys show that most people have thought of someone for no apparent reason, and then that person called, or else they just know who is calling before looking at the caller ID or answering the call. But couldn't this just be coincidence? We think about other people frequently; sometimes, by chance, somebody rings while we are thinking about them; we may imagine it is telepathy, but we forget the thousands of times we were wrong. Or when we know someone well, our familiarity with her routines and activities enables us to know when she is likely to ring, even though this knowledge may be unconscious.

I searched the scientific literature to find out if these reasonable possibilities were supported by any data or observations. I could find no research whatever on the subject. In science it is not enough to have a hypothesis. Hypotheses need testing to find out if they are supported or refuted by the evidence.

I therefore designed a simple procedure to test both the chance-coincidence hypothesis and the unconscious-knowledge hypothesis experimentally. I asked volunteer subjects for the names and telephone numbers of four people they knew well, friends or family members. The subjects were then filmed continuously throughout the period of the experiment alone in a room with a landline telephone, without a caller ID system. If there was a computer in the room, it was switched off, and the subjects had no cell phones with them. My research assistant or I selected one of the four callers at random by the throw of a dice. We called the selected person and asked him to phone the subject in the next couple of minutes. He did so. The subject's phone rang. Before answering it, she had to say to the camera who, out of the four possible callers, she felt was on the line. She could not have known by knowledge of the caller's habits and daily routines, because in this experiment the callers were randomly selected by the experimenter.

By guessing at random, subjects would have been right about one time in four, or 25 percent. In fact, in hundreds of trials the average hit rate was 45 percent, very significantly above the chance level. None of the subjects was right every time, but they were right much more than they should have been if the chance coincidence theory were true. (A video of one of these experiments made by a British TV channel is on my website.)

This above-chance effect has been replicated independently in telephone telepathy tests at universities in Germany and Holland. I have also obtained very similar results in experiments with text messages and SMS messages.

Skeptical reactions

Informed skeptics like Professor Chris French, the former editor of the British magazine *The Skeptic,* do not deny that there is experimental evidence that suggests psi phenomena are real, but say there is not yet enough to convince them. By contrast, dogmatic skeptics are generally ignorant of the evidence, as I have found in my many encounters with leaders of the organized skeptical movement, including James Randi and Richard Dawkins.

One of your own favorite sayings is that "Skepticism is a method, not a position." But you have not practiced what you preach. In my experience, you have been prejudiced and unscientific.

For example, in 2003, *USA Today* published an article about my book *The Sense of Being Stared At*, describing my research on telepathy and the sense of being stared at. As a prominent professional skeptic, you were asked for your comments. You were quoted as saying, "[Sheldrake] has never met a goofy idea he didn't like. The events Sheldrake describes don't require a theory and are perfectly explicable by normal means."

It takes years to do careful research and publish it in peer-reviewed journals. By contrast, it only takes a few minutes to make an evidence-free claim to a journalist. Dogmatic skepticism is easy.

As you will remember, I emailed you to ask what your normal explanations for my results were. You could not substantiate your very public claim in a newspaper read by millions of people, and admitted you had not even seen my book. I challenged you to an online debate. You accepted the challenge, but said you were too busy to look at the evidence, and promised you would "get to it soon." You didn't. So I am pleased that we are having this debate now.

In November 2005, you attacked me in your *Scientific American* "Skeptic" column in a piece called "Rupert's Resonance." You ridiculed the idea of morphic resonance and stated that I proposed a "universal life force," a phrase I have never used. You also referred to claims by fellow skeptics that my experimental work was flawed. These false claims had already been refuted in peer-reviewed journals, and even in the *Skeptical Inquirer*. I wrote a brief letter to *Scientific American* to set the record straight, but it was not published, nor even acknowledged.

In 2010, you contrasted skepticism with denialism, as in climate change denial, or Holocaust denial, or evolution denial, in your *New Scientist* article "Living in Denial: When a Sceptic Isn't a Sceptic": "When I call myself a skeptic, I mean I take a scientific approach to the evaluation of claims. ... A climate denier has a position staked out in advance, and sorts through the data employing 'confirmation bias'—the tendency to look for and find confirmatory evidence for pre-existing beliefs and ignore or dismiss the rest. ... Thus, one practical way to distinguish between a skeptic and a denier is the extent to which they are willing to update their positions in response to new information. Skeptics change their minds. Deniers just keep on denying."

In my experience, many crusading skeptics are deniers. They are in fact pseudoskeptics. You have behaved like a denier yourself. But I hope your belief in free inquiry will come out uppermost.

—*Rupert*

* * *

Michael Shermer Responds

Dear Rupert,

Our letters seem to be crossing not only in the email but in content as well. In your latest letter, you have nicely summarized the research you believe supports your hypothesis of morphic resonance and its various manifestations, such as that people know when they are being stared at from behind and that dogs know when their owners are coming home. For both of these sets of experiments and corresponding published papers, I outlined the numerous methodological shortcomings identified by scientists who have taken the time to examine carefully your data and, in some cases, even to try to replicate your experiments without success.

In the paper "The 'Psychic Pet' Phenomenon: A reply to Rupert Sheldrake" by Richard Wiseman, Mathew Smith, and Julie Milton, for example, the authors reveal what happens when you operationally define what constitutes a "hit" in psychical research: in this case, whether or not Jaytee the dog knew when his owner, Pam, was coming home. In initial observations by the owner, it seemed like Jaytee had foreknowledge based on his moving from the house to the porch. Sometimes Jaytee went to the porch when Pam was coming home, but there were plenty of times when Jaytee went to the porch with no connection to Pam at all. Wiseman *et al.* insisted on testing the claim by actually counting the number of such behaviors and especially the length of time Jaytee would stay on the porch waiting for Pam to return. Under these conditions, Jaytee went to the porch 12 times without correlation with Pam's return. One explanation for this nonsignificant finding is that Jaytee was distracted by the neighbor dog in heat and thus went to the porch with something else in mind. Wiseman et al. returned months later and carried out two more experiments, but failed to find any pattern between Jaytee's behaviors and that of his owner.

Your attempt, after the fact, to find a pattern in the video data by changing the criteria of a two-minute stay on the porch to 10-minute chunks of time—during which Jaytee allegedly spent more time on the porch during those periods when the owner was returning home than not—was gainsaid by the authors. They noted that such patterns should arise naturally by the fact that a dog is likely to do little after its owner departs, but then as the day goes on he is more likely to start anticipating the owner's return (just by normal time elapse and the dog's memory of the owner's usual time away) and make more trips to the porch. As well, searching the video record *post hoc* for patterns is a form of data snooping that is subject to the confirmation bias, and allowing Pam to determine when she would come home means her behavioral patterns might not be random at all but subject to her own unconscious preferences that Jaytee may have learned over time.

These particular methodological problems are not uncommon in psychical research, and thus my opinion of all such experiments is similar to that of the renowned paranormal researcher and experimental psychologist Susan Blackmore, who was once a believer in ESP but gave it up when she could not find enough convincing evidence: The tighter you make the controls and the more carefully you operationally define the behaviors to be measured, the weaker the psi effects become. You accuse me and other skeptics of being closed-minded materialists unable to see what is before our very eyes. Yet Blackmore set out on her professional career as a trained experimentalist and believer in psi to find evidence for the paranormal and came up empty handed, as she recalls in her essay, "Why I Had to Change My Mind," published in *Psychology: The Science of Mind and Behaviour, Sixth Edition,* by Richard Gross (Hodder Education, 2010):

> The results were a shock. Whether I looked for telepathy or precognition or clairvoyance, I got only chance results. I trained fellow students in imagery; chance results. I tested twins in pairs; chance results. I worked in play groups with very young children; chance results. I trained as a Tarot reader; chance results. Occasionally I got a significant result. Oh the excitement! Then as a scientist must, I repeated the experiment, checked for errors, redid the statistics, and varied the conditions, and every time either I found the error or got chance results again.

It takes considerable intellectual integrity to admit when your beliefs are wrong, but Blackmore has integrity in spades. As she explained about the day she became a skeptic: "At some point something snapped. Instead of struggling to fit my chance results into yet another doomed theory of the paranormal, I faced up to the awful possibility that I might have been wrong from the start—that perhaps there were no paranormal phenomena at all. I had to change my mind."

Rupert, the reason most of us scientists are skeptical of psi is not because we don't want it to be true; to the contrary. As I've explained in my earlier letters, if ESP, telepathy, telekinesis, or any other psi effects turned out to be real—and especially if an explanation for the effects were found (through neuroscience or quantum physics or whatever)—it would just be another remarkable feature of nature on the shelf next to the other natural wonders we already accept.

Like most dog owners, I would love to believe that I have a special psychical connection to my beloved chocolate lab Hitch (named for my friend, the late Christopher Hitchens) because I already know I have a physical and psychological bond with him and it would be easy to believe that there is something even more. But what I would like to be true and what is actually true may not always coincide (and he never seems to know when I'm going to crest the driveway hill on my bike ride home as he cluelessly snoozes on the porch). But the bond we have is beautiful and wonderful just for what it is, so I don't feel the need to believe there is more if there isn't.

And like so many viewers (I suspect), my wife and I were moved to tears—weeping really—during the Richard Gere film *Hachi: A Dog's Tale*, the true story of the faithful Japanese Akita Inu dog named Hachiko, who patiently waited for his owner Hidesaburo Ueno to return to the train station from his office—fruitlessly, every day for nine years—because his owner died at work. (Anyone who can get through this film without sobbing must be heartless.) If such connections as you suspect exist, why didn't Hachiko know that his owner was never coming home? If there is an afterlife, and the departed can communicate from beyond to those whom they love on this side, why didn't Hidesaburo give Hachiko some signal? Why allow such suffering to continue if we can do something about it? If love connects us, then why do so many people (and animals) in love not know when something tragic like this happens? We hear about the anomalous links between loved ones through powerful anecdotes that confirm such beliefs, but it is in the exceptions to the pattern that we must come to terms with and face the fact that love in this world is enough for you, for me, for Hitch, for Hachiko, and for everyone who has ever loved— and especially everyone who has loved and lost.

—*Michael*

* * *

Rupert Sheldrake Replies

Dear Michael,

Probably most of our readers will have experienced the sense of being stared at and telephone telepathy. Many will also have come across dogs or cats that know when their owners are coming home. Surveys in the U.S. and in Europe have shown that about 85 percent of people have experienced the sense of being stared at, about 80 percent of people have thought of someone for no apparent reason who then called, and about 50 percent of dog owners and 30 percent of cat owners say their pets anticipate the arrival of a member of the household. These are ordinary, not extraordinary, experiences— normal, not paranormal, events—and the scientific evidence supports their existence, as I discussed in my opening statement and in my response to yours.

Unfortunately, your dog Hitch seems to be in the 50 percent of dogs that don't sense when their owner is coming. The fact that the Japanese dog Hachiko, to whom you referred, waited every day at the train station for his deceased owner shows that this dog's devotion was not dimmed by his owner's death: this was also the case with a famous Scottish dog, Greyfriars Bobby, who spent 14 years guarding his owner's grave. But these animals' heroic loyalty is not a refutation of many dogs' and cats' abilities to know when their owners are coming home in normal circumstances.

Committed skeptics try to dismiss psi phenomena as tricks of the mind, mistaken interpretations of chance events, self-delusions, or examples of "anomalistic psychology." In effect, they are asking people to disregard their own experience in favor of the materialist theory that the mind is nothing but the brain, mental activity is nothing but brain activity, and minds are confined to the inside of heads. Therefore, mental action at a distance is impossible, or at least so unlikely as to merit no serious attention. Crusading skeptics also try to muddy the waters by making it sound as if positive results in psi research are false, or at least scientifically unreliable.

There is nothing wrong with fair criticism; science thrives on it. But science does not thrive on bias, prejudice, and willful attempts to cause confusion and misdirect attention. Professional skepticism is all too easy: Only the opinions of other skeptics count, and there is no need to do time-consuming experiments, or even to read about them. Moreover, skepticism pays, and it opens the way to a niche career in the media. When Susan Blackmore gave up her unsuccessful experimental research and joined the skeptic movement, her media career took off.

You and I have both referred to Richard Wiseman's claims to have debunked the "psychic pet" phenomenon. Below is my reply to his article, to which you referred our readers. In addition, I made a short film—*Jaytee, Pam Smart, Rupert Sheldrake, and Richard Wiseman: Setting the Record Straight*, which can be seen on YouTube and my website—that summarizes Wiseman's claims and shows how misleading they are. And Wiseman's theory that Jaytee went to the window more and more the longer Pam was out has already been refuted experimentally. You can see the data from the control experiments, filmed on occasions when Pam was not coming home, in Figure 3 in my summary of this long-running controversy, "Richard Wiseman's claim to have debunked the 'psychic pet phenomenon,'" which is also available on my website.

From the *Journal of the Society for Psychical Research* (2000), **64**, 126-128:

In the January issue of the *Journal* Richard Wiseman, Matthew Smith & Julie Milton published a reply to my note (Sheldrake, 1999a) about their claim to have refuted the "psychic pet" phenomenon. This claim was made in the *British Journal of Psychology* (Wiseman, Smith & Milton, 1998) and widely publicized in the media. It was repeated as recently as February 2 this year in a presentation given by the first author at the Royal Institution entitled "Investigating the Paranormal."

At my invitation, Wiseman and Smith carried out four videotaped experiments with a dog called Jaytee, with whom I have carried out more than 100 videotaped experiments (Sheldrake, 1999b). My experiments showed that Jaytee usually waited by the window for a far higher proportion of the time when his owner was coming home than when she was not. This occurred even when his owner, Pam Smart, came at non-routine, randomly-selected times and travelled by unfamiliar vehicles such as taxis. This pattern was already clearly apparent months before Wiseman *et al.* carried out their tests.

In the three experiments that Wiseman and Smith carried out at Pam's parents' flat, the pattern of results was very similar to my own. Their data show a large and statistically significant effect: Jaytee spent a far higher proportion of time at the window when Pam was on the way home than when she was not (Sheldrake, 1999a).

The difference between our interpretations of these experiments arose because Wiseman et al. had a different agendum from mine. I was engaged in a long-term study of this dog's anticipatory behaviour, whereas they seemed more interested in trying to debunk a "claim of the paranormal."

They themselves defined an arbitrary "claim" for Jaytee's "signal" and judged this by disregarding most of their own data. They argue that since they specified their criterion in advance (or rather criteria, since they changed the criterion as they went along), the agreement of their pattern of results with mine is irrelevant. "Testing this claim did not require plotting our data and looking for a pattern" (Wiseman, Smith & Milton, 2000, p. 46). Although thy refused to look at the pattern shown by their own data, I plotted their data for them, together with plots of my own data showing the same obvious pattern. I gave them these graphs before they submitted their paper to the British Journal of Psychology in 1996. Both in their paper and in their skeptical claims in the media, they chose to ignore what their own data showed.

"We tried the best we could to capture this ability and we didn't find any evidence to support it, " Smith was quoted as saying, in an article entitled "Psychic pets are exposed as a myth" (Irwin, 1998). "A lot of people think their pet might have psychic abilities but when we put it to the test, what's going on is normal not paranormal," Wiseman asserted in the press release accompanying their paper. These are examples of the comments they now describe as "responsible and accurate."

Wiseman, Smith & Milton try to justify ignoring the pattern shown by their data on the grounds that it was "post hoc." I cannot accept this argument. First, I had been plotting data on graphs right from the beginning of my research with Jaytee. Second, their dismissal of post hoc analysis would deny the validity of any independent evaluation of any published data. The whole point of publishing scientific data is to enable other people to examine and analyze them. Of necessity, the critical analysis of published data in any field of research can only be post hoc. And third, the plotting of graphs is not normally regarded as a controversial procedure in science. Consequently, I do not agree with them that my representation of their results in my book (Sheldrake, 1999, Figure 2.5) is "misleading."

In their recent note, they raise two scientific, as opposed to legalistic, points. First, they suggest that Jaytee may simply have gone to the window more and more the longer Pam was out, and hence been there most when she was on the way home. But a comparison of Jaytee's behaviour during Pam's short, medium and long absences shows that this was not the case (Sheldrake, 1999b). Moreover, in control experiments in which Pam did not come home, Jaytee did not go to the window more and more as time went on (Sheldrake, 1999b, Figure B.2).

Second, they say that my experiments "appear to contain design problems (Blackmore, 1999)." Susan Blackmore's comments were made in an article in the *Times Higher Education Supplement*, which concluded: "There are better ways to spend precious research time than chasing after something that lots of people want to be true, but almost certainly is not." She thought she had spotted "design problems" in my experiments with Jaytee (Sheldrake 1999b) because "Pam was never away for less than an hour." (In Wiseman, Smith & Milton's experiments Pam was likewise never away for less than an hour.)

This is why Blackmore thought there was a problem: "Sheldrake did 12 experiments in which he bleeped Pam at random times to tell her to return..... When Pam first leaves, Jaytee settles down and does not bother to go to the window. The longer she is away, the more often he goes to look.

[Y]et the comparison is made with the early period when the dog rarely gets up." But anybody who looks at the actual data (Sheldrake, 1999b, Figure B.4) can see for themselves that this is not true. In five out of the 12 experiments, Jaytee did not settle down immediately she left. In fact he went to the window more in the first hour than during the rest of Pam's absence, right up until she was on the way home, or just about to leave.

In the light of Blackmore's comments, I have reanalyzed the data from all 12 experiments excluding the first hour. The percentage of time that Jaytee spent by the window in the main period of Pam's absence was actually lower when the first hour was excluded (3.1%) than when it was included (3.7%). By contrast, Jaytee was at the window 55.2% of the time when she was on the way home. Taking Blackmore's objection into account strengthens rather than weakens the evidence for Jaytee knowing when his owner was coming home, and increases the statistical significance of the comparison. (Including the first 60 minutes of Pam's absence in the analysis, by the paired-sample t test, $t=-5.72$, $p=0.0001$; excluding the first 60 minutes, $t=-5.99$, $p<0.0001$.)

Blackmore's claim illustrates once again the need to treat what sceptics say with scepticism.

In conclusion, I agree with Wiseman, Smith & Milton (2000, p.49) that my analysis of their data "would not provide compelling evidence of psi ability unless it were supported by a larger body of research." It is in fact supported by a large body of research, summarized in my book (Sheldrake, 1999b) and soon to be published in detail in a peer-reviewed journal.

Update May 1, 2007 - Skeptic demonstrates psi
Richard Wiseman admitted in his recent *Skeptiko* interview that his data does correspond with Sheldrake's.

I wish there was a way to move our argument forward. It seems to me that your rejection of psi phenomena is a result of your worldview and confirmation bias. Based on my experience of our dialogues so far, I do not expect that anything I write, or any evidence I draw to your attention, will lead you to change your beliefs. But I live in hope. I do not know what you will say in your reply, published simultaneously with this. I will read it with your own criterion in mind: "One practical way to distinguish between a skeptic and a denier is the extent to which they are willing to update their positions in response to new information. Skeptics change their minds. Deniers just keep on denying."

Some of our readers may want to find out more about mental action at a distance. They do not need to take my word for it, or yours against it. I encourage those who are interested to discuss their own experiences with their friends and families, to read some of the many published papers on psi research (available on the website OpenScience.org), and to try some experiments for themselves. On my website, you can find two experiments that can be carried out with one or two other people online or via phone.

—Rupert

PART 3
GOD AND SCIENCE

Dr. Sheldrake defends the position that there is no conflict between science and the existence of God; evidence from conscious experience renders belief in God reasonable.

Dr. Shermer opposes Dr. Sheldrake's position, arguing that science in no way supports the existence of God; in fact, science undercuts the reasonableness of belief in God.

Chapter 5

Rupert Sheldrake's Opening Statement

Dear Michael,

I believe in God, and I am a church-going Christian, an Anglican (Episcopalian in the United States). You are a materialist, atheist, and secular humanist. I think faith in God and the practice of science are compatible and mutually enhancing. I dare say you disagree.

Nevertheless, we probably agree about many things. First, we both believe that the universe has a unity that makes science possible. This belief is shared by Christians, atheists, and by followers of other faiths. Second, we both agree that the universe is intelligible, at least in part—otherwise, science and reason would be futile.

For those who believe in God, the intelligibility of nature and the ability of human minds to understand some aspects of the natural world make sense because they have a common source, namely God. God's consciousness is the ultimate source of human consciousness, and all other forms of consciousness in the universe.

Most atheists and materialists also believe in something like the mind of God, but stripped down to mathematical principles, or "the laws of physics," which were there from the beginning, and are changeless, universal, and omnipotent. This seems to be the opinion of your friend Lawrence Krauss, for example. The main difference is that in Christian theology, the ordering principles of the universe are aspects of God, whereas for an atheist the laws of nature sprang into being miraculously at the moment of the Big Bang, or exist in a transcendent Platonic realm, from which they somehow gave rise to the universe.

If these laws are explained in terms of yet more fundamental laws, as in M-theory, or superstring theory, then where do those ultimate laws come from? Just like the woman who thought the world rested on a turtle, and that turtle on another turtle, and so on all the way down, in modern physics, mathematical laws rest on mathematical laws all the way down.

You will argue that to say that the ordering principles of nature have their ultimate source in God adds nothing simply to saying there are laws. But it does add something. God is by definition a conscious being, and divine consciousness permeates the universe, and is also the ultimate source of human consciousness. By contrast, materialists believe that the universe is unconscious, governed by unconscious laws, and made up of unconscious matter. These assumptions make human consciousness problematic, which is why philosophers of mind call the very existence of consciousness "the hard problem."

A third area in which we agree is evolution. We both believe in evolution, not only at the biological level, but also at the cosmic level. The whole universe has been expanding and developing for billions of years, forming ever more varied structures within it, including galaxies, solar systems, and biological life, at least on this planet. All the new forms that come into being, including the forms of molecules, crystals, plants, and animals, require some kind of creativity. Like you, I think that creativity is inherent in nature, and I do not think that the universe is designed by an external engineering-type God. Also, like you, I do not think that the Bible, or any other sacred book, is a scientific, historical record to be taken literally.

Some biblical fundamentalists think of God as an engineer who designed and created species of animals and plants like a watchmaker designing a watch. Ironically, this God of the world machine has more to do with science than with the Bible or traditional Christian doctrines. When the machine model of nature took hold in 17th-century science, a new image of God came into being as a supernatural engineer, a machine-maker separate from nature.

You don't believe in this kind of God, and neither do I. In traditional Christian theology, God is not a kind of craftsman, or demiurge, who makes the world in the first place and then retires, leaving it to work automatically, except for occasional interventions when he arbitrarily suspends the laws of nature. God is not a demiurge, and not a meddler with machinery. According to the traditional understanding in Christian and other theologies, God is the ground of all being, the reason why there is something rather than nothing. He sustains the world in its existence from moment to moment, and is doing so now.[1]

In any case, the biblical account of creation does not see nature as mechanical, or God as an engineer.

Plants and animals were not invented by God in a kind of celestial workshop. The Creation account in the book of Genesis (1.11) reads, "Then God said, 'Let the earth put forth vegetation: plants yielding seed and fruit trees of every kind on earth that bear fruit with the seed in it.' And it was so."[2] Likewise, God empowered the seas and the earth to bring forth marine, terrestrial, and flying animals (Genesis 1.20–24). In theology this is called "mediate creation." God did not create plants and animals directly, micromanaging their details, but endowed Nature with an inherent creativity.

Spiritual traditions in general, and religions in particular, were not founded on irrational propositions, or on blind faith, or on dogma, or on fear. They arose from states of consciousness that go beyond normal everyday experience. Shamans, Indian *rishis*, the Buddha, the Jewish prophets, Jesus, and Muhammad spoke from their direct experiences of connection to a greater consciousness.

In your opening statement on Materialism in Science back in May, you remarked: "If God (or some creative force—it need not be the creator Judeo-Christian-Muslim God) exists outside of nature, but periodically reaches into our world to change it in some manner ... then, in principle, there should be some way to measure the effect."

I do not believe that God "periodically reaches in." He is continuously present throughout all of nature. But if you want evidence of the effects of God, or a creative force, then look at the way great religious leaders have changed the course of human evolution; if you want material evidence, then look at temples, cathedrals, and mosques.

Throughout the history of humanity, some people have connected with realms of consciousness beyond the human level, and many still do so today. Some people have spontaneous mystical experiences; some have their minds opened by psychedelics; some connect with God, or saints, or angels, or ancestors, through prayer; some meditate; some connect to spiritual realms in other ways. The British biologist Sir Alister Hardy, who studied the natural history of religious experiences in the modern world, found that they were far more common than most people assumed.[3]

Not all religious people have had such experiences themselves, just as not all people who believe in a scientific worldview have experienced working in a research laboratory or in an institute of theoretical physics. Many religious and scientific beliefs are accepted secondhand or thirdhand on the basis of religious or scientific authority.

I agree with you that the direct experience of divine presence is not a conclusive argument for the existence of God, however convincing it seems to people who experience it. After all, some atheists who have taken psychedelics remain atheists, and see their mind-expanding experiences as nothing more than chemical disturbances of normal brain functioning, rather than opening the doors of perception. Anti-psychedelic religious people agree. Some atheists practice meditation and experience expanded states of mind while remaining atheists, including Susan Blackmore and Sam Harris. They—and probably you, too—would no doubt explain these experiences as being produced inside the brain as a result of physiological changes in the nervous system, proving nothing about consciousness beyond human brains. But this interpretation is itself a product of belief. If you are committed to the materialist worldview, it is an article of faith that the mind is confined to the brain and cannot connect with a greater mind that pervades the universe, because such a mind does not exist.

Our differences are not simply about beliefs. These beliefs have effects in practice and affect the way we lead our lives. The materialist doctrine that all conscious experience is nothing but the activity of the brain has an isolating effect, pulling people back into our own separate skulls. By contrast, for those who accept the reality of consciousness beyond our own, such experiences open channels of communication that can be pursued through meditation, prayer, rituals, festivals, worship, and thanksgiving—through many of the traditional practices of religion.

This experience of connection with the spiritual realm affects people's physical and mental health, as well. Numerous studies in the United States and elsewhere have shown that people who are religious, especially those who regularly attend religious services, live significantly longer, enjoy better health, and suffer less depression than people without religious practices. Both Christian and non-Christian groups showed these effects.[4]

In other studies, people who prayed or meditated were compared with similar people who did not. These studies were prospective, as opposed to retrospective: the people under observation were identified at the start of the study, and then watched over a period of years to see if their health and mortality turned out differently. They did. On average those who prayed or meditated remained healthier and survived longer than those who did not. In one study in North Carolina, Harold Koenig and his colleagues tracked 1,793 subjects over 65 years old who had no physical impairments at the beginning of the study. Six years later, those who prayed survived 66 percent more than those who did not pray.[5]

In our first dialogue, you wrote, "I'm encouraged by the recent increase in the number of people with no religious affiliation." From the point of view of evangelical atheism, this must be an encouraging sign. But will it benefit those who have abandoned religion? Some people brought up in a religious atmosphere find atheism liberating, at least for a while. I did myself, and for more than 10 years, I identified myself as an atheist. But I then came to see the atheist worldview as narrowly dogmatic, especially when it denied the value of spiritual experiences that I found enriching and enlivening. Moving beyond atheism to an acceptance of the spiritual realm felt like leaving a two-dimensional, black-and-white intellectual world for a full-color, three-dimensional reality.

If religious practices can lead to better health, then, as a corollary, the loss of faith and the cessation of religious practices can damage health and well-being. Atheism is not just about intellectual theories and the denunciation of religion. It can have serious adverse effects. Like smoking, it should carry a health warning.

The British philosopher Alain de Botton, a second-generation atheist, thinks much the same. In his bestselling book *Religion for Atheists* (Vintage, 2013), he argues that an atheist or agnostic lifestyle is severely impoverished, and suggests that non-believers should learn from religions "how to build a sense of community, make our relationships last, overcome feelings of envy and inadequacy, escape from the 24-hour media world and get more out of art, literature, and music." His practical suggestions include atheist festivals, atheist temples, atheist sermons, and atheist Sunday assemblies. I agree with his analysis of the needs unsatisfied by atheism and agnosticism, but see no need to reinvent religion when we have plenty already and are spoilt for choice.

Michael, I agree with you that some religious people have done terrible things in the past, and some still do so. A notorious example was the Spanish Inquisition, which operated from 1478 to 1834 and was responsible for as many as 5,000 executions.

Many anti-religious people have also behaved murderously, including Stalin, Mao Zedong, and Pol Pot, all of whom were atheists and believers in science and reason, following Karl Marx. Together, these communist regimes led to at least 20 million deaths. I know that you are trying to rebrand communism as a "faux religion" to avoid acknowledging that atheists have caused human misery on a vast scale. But the fact is that some people are capable of doing very bad things to other people using religion, atheism, nationalism, ideology, profit, or even reason as their justification. Reason was a major inspiration for the French Revolution, and in 1793, the Cathedral of Notre Dame in Paris was converted into a Temple of Reason, and the Cult of Reason was proclaimed the state religion. At the same time, at least 40,000 people were executed in the Reign of Terror (1793–4), and the guillotine became a symbol of the revolutionary cause.

Finally, you and I probably agree that a great deal of religion is culturally determined and subject to human limitations. You may see this as another refutation of religion, but I see it differently. The core of all religions is the experience of connecting with the ground of being, or the mind of God, or ultimate consciousness, but when these experiences are talked about and interpreted, they are refracted through human languages and cultural traditions that are necessarily limited, and inevitably different from one another.

If I had been born to Sinhalese parents in Sri Lanka, I would probably be a Buddhist today. If I had been born to Muslim parents in Egypt, I would probably be a Muslim. In fact, I was born to Christian parents in England, and the form of religion that I find most congenial is the Christian faith, and in particular the Anglican form of Christianity. This does not mean I believe other Christian churches are in error, and that other religions are wrong. It does not mean that I think that people who are spiritual but not religious are lost. There are different paths to God, or to ultimate reality, and all have their strengths and weaknesses.

I see faith in God and the practice of science as complementary, not contradictory. Conflicts arise from dogmas on both sides. Religious and scientific fundamentalists are still locked in old battles, and some people on both sides relish the fighting. Fortunately, the sciences themselves are moving beyond the dogmas of materialism, and new possibilities for dialogue between the sciences and religious traditions are opening up.

—*Rupert*

* * *

Michael Shermer Responds

Dear Rupert,

You are correct, I do not believe that "faith in God and the practice of science are compatible and mutually enhancing." In fact, when phrased this way they are completely different things: non-overlapping magisteria (NOMA) in the memorable description of my late friend Stephen Jay Gould. In no way is science based on faith—and this fact puts the lie to the fatuous claim by anti-scientism folks that science is a religion, or that it is just another way of knowing, no better than any other. Faith, in the biblical meaning, has nothing to do with science. It is, as famously defined in Hebrews 11:1, "the assurance of things hoped for, the conviction of things not seen."

Science rests on the assurance of things predicted, the conviction of things previously seen. Although scientists, being human, may hope that their unseen convictions are supported by the data, as a system of knowledge the practice of science has more in common with plumbing than it does religion, in the sense that at least plumbers test hypotheses when they attempt to discover and fix leaky pipes, clogged toilets, plugged sinks, and the like. Reason, logic, empiricism, mathematics, and experimentation—the core tools in the practice of science—are the exact opposite of faith in the biblical sense. Having faith in things is most assuredly a human trait, but it is not a part of the scientific method because it is not a reliable epistemology.

You assert: "God's consciousness is the ultimate source of human consciousness." Where is the evidence for this claim? As I challenged you previously with regard to the resurrection of Jesus, where's your control group? In any case, if human consciousness comes from God's consciousness, what is the source of God's consciousness? If you argue that you have to stop the causal chain somewhere, why stop at God? If there's a chain of being from lower to higher consciousness, why would there not be an über-God consciousness? If you believe that consciousness is a property of the universe (which you seem to argue in your theory of morphic resonance), then why do you need God at all? It seems entirely reasonable to me to argue (along these lines) that if consciousness exists itself separate from any entities, then there is no need for God. If God is necessary for the existence of consciousness, then once again I would challenge you to go one step further in inquiring into the source of God and God's consciousness.

You ask where the laws of nature come from without postulating God as the source; that is, in a purely materialistic naturalistic worldview. As you know, this is one of the biggest questions in the philosophy of science. No one knows for sure, but I am encouraged by the work of Stephen Hawking and Leonard Mlodinow, outlined in their book *The Grand Design* (Bantam, 2010), that certain configurations of the laws of nature inevitably lead to the spontaneous creation of universes. They show how the laws underlying quantum mechanics and relativity, for example, could lead to universes being formed out of nothing. They write:

> Because there is a law such as gravity, the universe can and will create itself from nothing. Spontaneous

creation is the reason there is something rather than nothing, why the universe exists, why we exist. It is not necessary to invoke God to light the blue touch paper and set the universe going.

Another explanation is the multiverse theory. In his book *God: The Failed Hypothesis* (Prometheus Books, 2007), the late physicist Victor Stenger shows how there could be as many as 10^{500} possible universes. Those universes with the laws of nature that lead to the production of atoms and molecules and life and sentient beings who inquire as to their source, will naturally seem almost miraculously designed by those fortunate sentient beings, but with that many universes there was bound to be some configured purely by chance to be conducive to the evolution of life and sentient beings, but there's nothing designed about it.

In any case, science is young—just a few centuries old—and there is much we do not know, so before we speculate about such ultimate questions as where the laws of nature came from, let's wait to see what science comes up with in the next couple of centuries before we postulate a divine being as the source. In the meantime, it's okay to say "I don't know" and leave it at that. Humans have always filled in such gaps in our knowledge with gods, and it never leads to any useful or productive theory. Let's try to overcome this psychological propensity to fill in the gaps with supernatural forces and follow the path of science in searching for natural forces.

You write: "God is by definition a conscious being." How do you know? Whose definition? Sources and evidence, please.

You claim that "divine consciousness permeates the universe" and that we materialists "believe that the universe is unconscious, governed by unconscious laws, and made up of unconscious matter." First, where is divine consciousness in the universe? In what way are, say, stars conscious? Stars are massive collections of hydrogen atoms being converted to helium atoms through nuclear fusion. In what way is this process "conscious"? Are galaxies conscious? Is gravity conscious? Are planets, moons, and comets conscious? What do you mean by consciousness here? Clearly this has no resemblance to human consciousness—the phenomenon of being aware and self-aware—so it would be helpful to operationally define consciousness.

Your idea about what God is seems to differ dramatically from that of most Christians (a cosmic engineer or divine craftsman), and you reject the idea that God reaches into the world to stir the particles now and then. Instead, you invoke the unhelpful phrase "ground of being" as your definition of God, and you claim that "he sustains the world in its existence from moment to moment, and is doing so now." I presume you mean that God does so through the laws of nature—he forms solar systems through the force of gravity, he forms stars through the nuclear forces, etc. I fail to see how this is any different from just saying that the universe itself is the "ground of being" and that it sustains itself from moment to moment through the laws of nature.

What's the difference between an invisible God that is indistinguishable from nature and a nonexistent God?

—*Michael*

* * *

Rupert Sheldrake Replies

Dear Michael,

You believe that an unconscious, purposeless universe produced minds in human brains after about 14 billion years of mindless mechanical activity. You trust that science will eventually justify this belief. I believe that consciousness comes first, and is the source of nature and of minds within nature.

You ask me how I know that God is conscious, and ask for "sources and evidence." There are many religious sources. In the Old Testament, for example, in Exodus 3:14, Moses asks God for his name and God replies, "I am who I am" (English Standard Version). God defines himself as subjective conscious being, in the present. As I mentioned previously, one of the Indian names for ultimate reality is *satchitananda*, being-consciousness-bliss, again emphasizing conscious being. Evidence comes from mystical experiences and through moments of illumination, in which people feel themselves in the presence of a greater consciousness than their own. Evidence of consciousness comes through conscious experiences, not from physical measurements.

You propose that the source and sustainer of all things is not God, but the "laws of nature." However, this conception of natural laws is not as atheistic as it seems, because it builds on the belief that the laws of nature are aspects of the mind of God, a belief shared by the founding fathers of modern science in the 17th century, including Galileo, Kepler, Descartes, and Newton. But laws are ideas rather than material objects, and they make no sense without a mind to think them.

You endorse the speculations of Stephen Hawking and Leonard Mlodinow that "Certain configurations of the laws of nature inevitably lead to the spontaneous creation of universes." Somehow, these laws "could lead to universes being formed out of nothing." This is again very like 17th-century theology, but less coherent, because you imagine that timeless laws exist in a free-floating, transcendent realm—God's mind without God—and generate a universe without a source of energy or activity. In the 17th century, God was seen as the source of all activity, as well as the basis of cosmic law and order.

Nevertheless, you often contradict yourself. In our first dialogue you wrote, "'Laws' are the linguistic and mathematical descriptions we humans give to naturally occurring repeating phenomena. There are no laws of nature 'out there.'" But if they are merely descriptions of the regularities of nature, how can laws precede the universe and give rise to it? Before the origin of the universe there were no regularities of nature and no humans to describe them.

And why are these hypothetical laws just right for our universe, enabling it to produce life and ultimately humans? Here, you fall back on the speculations of Victor Stenger, who argues that we happen to be in the only universe just right for us, out of 10^{500} universes. He and other cosmologists avoid awkward problems by quantitative easing, conjuring up myriad universes, just as the Federal Reserve Bank conjured up trillions of dollars. It may be easier to proliferate universes than to reconsider the existence of God. But, as philosophers point out, an infinite God could be the God of an infinite number of universes. The multiverse gambit fails to get rid of God, and it is the ultimate violation of Ockham's Razor, the principle that "entities should not be multiplied unnecessarily."

You contradict yourself again. In some paragraphs, you see science "as the conviction of things previously seen," and the practice of science like plumbing, fixing leaky pipes and clogged toilets, based on hard facts. You see this as "the exact opposite of faith in the biblical sense." But a belief in 10^500 unobserved universes is not a hard fact. Nor is promissory materialism. If science cannot explain something, no problem! Just give us time, you say, "Let's wait to see what science comes up with in the next couple of centuries before we postulate a divine being as the source." This is faith as described in Hebrews 11:1, "the assurance of things hoped for, the conviction of things not seen."

I think God is within nature, and nature within God. You think the universe is purposeless and unconscious, and that it originated from mindless yet immaterial laws.

For me, God comes before humanity; for you, as a secular humanist, humanity comes before God, who is a delusion in human minds, and hence in human brains.

Our beliefs affect our lives. No one can wait 200 years to make a choice.

—*Rupert*

Chapter 6

Michael Shermer's Opening Statement

Dear Rupert,

For our final three rounds on the existence (or not) of God, I will be especially curious to know how you arrive at orthodox Christianity out of your unorthodox theory. As you told my *Scientific American* colleague John Horgan:

> I believe in God. I am a practicing Christian, specifically an Anglican (in the U.S.,an Episcopalian). I went through a long atheist phase, and began to question the materialist orthodoxy of science while I was still an atheist. I later came to the conclusion that there are more inclusive forms of consciousness in the universe than human minds.

Already, I'm suspicious—and skeptical. How convenient that the God you believe in happens to be the same God that most of your fellow countrymen (and, more broadly, your fellow Westerners) believe in. Geography, in fact, is the number one predictor of anyone's religion—where they happen to have been born and raised. Had you been born and raised in, say,

India, it would be far more likely that you would defend Hinduism as the One True Religion, the god Ganesha as your deity of choice, and employ morphic resonance theory to explain reincarnation and Karma (memory of past lives and deeds).

But let me not presume too much, and await your statement—and in the meantime, inquire why belief in the supernatural (as you have defined and defended it) and belief in God should be conflated at all. Historically speaking, most commentators have lumped the two together, but it seems to me entirely possible that the two could be separate. Although I do not believe this myself, assuming for a moment that you are correct that there is a world beyond the natural world (or, if you agree with me that there is only the natural word, then forces heretofore unaccepted by mainstream science such as psi really do exist), such a world could contain no God, one God, or a multitude of Gods. In principle, psi-like forces could exist whether or not there's a God. You could be a supernatural atheist in this sense.

Likewise, there could be a God and no supernatural forces. God could just use the known forces of nature (or perhaps some heretofore unknown forces of nature) to perform miracles. And if there is a God, why not multiple gods? There is nothing inherently special about monotheism, save for its historical triumph as a religion—perhaps superior at uniting tribes against other tribes or playing a prominent role in the first civilizations to reinforce group cohesion and act as an enforcer of morals and ultimate punisher of violators. This is, in fact, my theory of the origin of religion (developed in my books *How We Believe* [W.H.

Freeman & Co., 1999] and *The Believing Brain* [Times Books, 2011]), which I define as a social institution to create and promote myths, to encourage conformity and altruism, and to signal the level of commitment to cooperate and reciprocate among members of a community. There are multiple lines of evidence that humans created gods, not *vice versa*.

First, there are many human universals that anthropologists have discovered among all peoples of the world related to religion and belief in deities, such as sacraments surrounding death, supernatural beliefs about fortune and misfortune, and especially divination, folklore, magic, myths, and rituals. With such universals we can presume that there is a genetic predisposition for these traits to be expressed within their respective cultures, and that these cultures, despite their considerable diversity and variance, nurture these natures in a consistent fashion toward belief in gods and religious rituals.

Second, twin studies have consistently found a strong genetic component to religiosity and belief in God—roughly speaking, about 40 percent to 50 percent of the variance is accounted for by genetics. Genes, of course, do not make one a Jew, Catholic, Muslim, or any other religion. Rather, genes code for cognitive and behavioral tendencies that make one more or less likely to believe in supernatural agents (God, angels, demons) and more or less likely to commit to certain religious practices (church attendance, prayer, rituals).

Third, the comparative study of religions leads to this back-of-the-envelope calculation: over the past 10,000 years of history humans have created about 10,000 different religions and about 1,000 gods. What is the probability that your God, Yahweh, is the One True God, and Amon Ra, Aphrodite, Apollo, Baal, Brahma, Ganesha, Isis, Mithras, Osiris, Shiva, Thor, Vishnu, Wotan, Zeus, and the other 986 gods are false gods? As skeptics like to say, everyone is an atheist about these gods—some of us just go one god further.

Fourth, since you are a Christian, I must point out that even within the three great Abrahamic religions there is much disagreement. Christians believe Jesus is the savior and that you must accept him to receive eternal life in heaven. Yet, both Jews and Muslims reject this central tenet of your religion. In fact, only roughly two billion of the world's six billion believers accept Jesus as their personal savior. Most Christians believe that the Bible is the inerrant gospel handed down from the deity. Muslims believe that the Koran is the perfect word of God, and yet Christians do not. As well, while most Christians believe that Jesus was the last prophet, Muslims believe that Muhammad is the last prophet, while Mormons believe that Joseph Smith is the last prophet. Who's to say which one is right? By what criterion is one to judge among these competing claims? And while they are usually called "faith-based" beliefs, believers in fact hold these tenets of their religion to be really true—actually and factually correct—not just relatively or psychologically true. Given that fact, Rupert, how do you decide which among the many world's religions is the right one? You're a scientist—if you believe Jesus was resurrected from the dead three days after being crucified, where's your control group?

Fifth, the study of comparative mythology leads to the recognition of commonalities among flood myths, virgin birth myths, and resurrection myths, all of which converge to the strong inference that they are not unique to your religion but were—like God—invented by humans.

Flood myths: The Noachian flood story is predated by the Epic of Gilgamesh by about four centuries. Warned by the Babylonian Earth-god Ea that other gods were about to destroy all life by a flood, Utnapishtim was instructed to build an ark in the form of a cube 120 cubits (180 feet) on all sides, into which he was instructed by god to put one pair of each living creature.

Virgin birth myths: Among those conceived sans a biological father were Dionysus, Perseus, Buddha, Attis, Krishna, Horus, Mercury, Romulus, and, of course, Jesus. The Greek God Dionysus, for example, was born from a virgin mortal mother impregnated by the king of heaven. He was said to be able to transform water into wine (he was, after all, the Greek god of wine), and it was he who introduced the idea of eating and drinking the flesh and blood of the creator, not the Catholics centuries later.

Resurrection myths: The ancient Egyptian god of life, death, and fertility, Osiris, first appears in the pyramid texts around 2400 B.C. Widely worshipped until the compulsory repression of pagan religions in the early Christian era, Osiris was not only the redeemer and merciful judge of the dead in the afterlife, he was also linked to fertility, most notably the flooding of the Nile and growth of crops. The kings of Egypt themselves were inextricably connected with Osiris in death, such that when Osiris rose from the dead, so would they in union with him. By the time of the New Kingdom, not only pharaohs but common people believed that they could be resurrected by and with Osiris at death—if, of course, they practiced the correct religious rituals.

Since you are a Christian, I presume that you believe God to be *all-powerful (omnipotent), all knowing (omniscient), and all good (omnibenevolent); who created out of nothing the universe and everything in it; who is uncreated and eternal, a non-corporeal spirit who created, loves, and can grant eternal life to humans.* I do not believe in this God. That makes me an atheist. Yes, I know, technically speaking, I cannot prove a negative—I cannot prove God does not exist. But neither can I prove that Zeus does not exist, and yet I don't believe in him, either. And as I showed above, I can provide strong evidence indicating that it is very likely God does not exist except in the minds of people. (That God is very powerful indeed, as he can get people to fly planes into buildings and to blow themselves up in crowded squares.)

As well, I can show that the traditional arguments for God's existence are refutable—and have been refuted, starting with Hume's devastating critiques. The first cause, prime mover, and cosmological arguments, the argument from the fine-tuning of the universe, the argument for the origin and intelligent design of life, the moral origin argument, the origin of consciousness argument, and the like, all have equally plausible (and in most cases more probable) explanations from science, and in any case constitute what is called the "god of the gaps" style of reasoning—anything that science cannot currently explain (a gap in our knowledge) is best filled with God as the explanation.

There are several fallacies with this line of reasoning. First, the either/or fallacy, or the false dilemma, is the tendency to dichotomize the world such that if you discredit one position it means the other position must be correct. Not so. It is not enough to call out the weaknesses in a theory; you must also provide evidence that your alternative theory is superior—that is, it explains more data than the alternative.

Second, science is young and there is much we still do not know. Before we say something is out of this world, let's first make sure that it is not in this world. We have some cogent natural explanations for the origin of the universe, life, consciousness, and morality, but much remains unknown.

Third, what will happen to the God hypothesis that relies on these gaps when the gaps are filled by science?

Finally, there is one God I could believe in based on a purely naturalistic worldview. In honor of Arthur C. Clarke and his famous three laws, I have called this Shermer's Last Law: any sufficiently advanced Extra-Terrestrial Intelligence (ETI) is indistinguishable from God. (Clarke's Third Law states: "Any sufficiently advanced technology is indistinguishable from magic.")

My gambit is based on three observations and three deductions:

- *Observation I.* Biological evolution is glacially slow compared to cultural and technological evolution.
- *Observation II.* The cosmos is very big and space is very empty, so the probability of making contact with an ETI is remote.
- *Deduction I.* The probability of making contact with an ETI who is only slightly more or less advanced than us is virtually nil. Any ETIs we

would encounter will either be way behind us
or way ahead of us.

- *Observation III.* Science and technology have
changed our world more in the past century
than in the previous hundred centuries.
Moore's Law of computer power doubling
every 12 months applies to dozens of other
technologies. If Ray Kurzweil is right in his
book *The Singularity is Near* (Viking, 2005), then
the world will change more in the next century
than it has in the previous thousand centuries.
- *Deduction II.* Extrapolate these trend lines out
tens of thousands, hundreds of thousands, or
even millions of years—mere eye blinks on an
evolutionary time scale—and we arrive at a
realistic estimate of how far advanced an ETI
will be.
- *Deduction III.* If today we can engineer genes,
clone mammals, and manipulate stem cells with
science and technologies developed in only the
last 50 years, think of what an ETI could do with
50,000 years of equivalent powers of progress
in science and technology. For an ETI who is
a million years more advanced than we are,
engineering the creation of planets and stars
may be entirely possible. And if universes are
created out of collapsing black holes—which
some cosmologists think is probable—it is not
inconceivable that a sufficiently advanced ETI
could even create a universe by triggering the
collapse of star into a black hole.

What would we call an intelligent being capable of engineering life, planets, stars, and even universes? If we knew the underlying science and technology used to do the engineering, we would call it an Extra-Terrestrial Intelligence; if we did not know the underlying science and technology, we would call it God.

Either that God already exists, or we are becoming that God. Either way, the mere contemplation of the possibilities simply blows one's mind. If you want to be awestruck by a religious-like experience, turn to science.

—*Michael*

* * *

Rupert Sheldrake Responds

Dear Michael,

Our most fundamental disagreement is about consciousness beyond the human level. As a materialist and atheist, you regard it as impossible or at least highly unlikely, unless it takes the form of Extra-Terrestrial intelligence in a civilization with technologies far more advanced than our own. What you call "Shermer's Last Law" is that "any sufficiently advanced Extra-Terrestrial Intelligence is indistinguishable from God." You imagine humanoids that have become omnipotent through science and technology "engineering the creation of planets and stars" and "even creating a universe by triggering the collapse of a star into a black hole." And after developing your fantasy scenario, you conclude, "If you want to be awestruck by a religious-like experience, turn to science." But this is science fiction, not science.

Nevertheless, we have several areas of agreement. Like you, I do not think that psychic phenomena such as telepathy are evidence for God or a supernatural realm. As I pointed out in our last dialogue, I see psi phenomena as part of the natural world; they have evolved in many species of social animals. The existence or non-existence of God is a separate question. Many psi researchers and religious people agree with us. Several leading parapsychologists are atheists: They accept the existence of psi but not of God. Meanwhile, some believers in God disbelieve in psi phenomena, or disapprove of them. One of the founders of organized skepticism in the U.S., Martin Gardner, was a theist who was vehemently opposed to research in parapsychology because he thought it was "tempting God" and seeking "signs and wonders."

We also agree that there are similar features in different religious traditions, including stories of floods, virgin births, and resurrections. I see these themes as archetypal, reflecting fundamental ways of understanding the world, or as cultural memories. Flood myths may well be related to the actual floods that happened at the end of the last ice age when sea levels rose dramatically. One aspect of virgin birth myths is their emphasis on the creative power of the life-giving mother. No doubt the cult of the Blessed Virgin Mary in the Roman Catholic and Orthodox churches inherits aspects of pre-Christian goddess worship, in which the Great Mother gave birth to gods. I do not see this as a problem. In fact, I see it as a strength of the Christian tradition that it includes pre-Christian elements such as pilgrimages and seasonal festivals. Many Christian sacred places were sacred to older religions first: For example, the shrine of Our Lady of Guadalupe in Mexico was built over the temple of the Aztec mother goddess. Religions evolve, like everything else.

The theme of death and resurrection is common to many traditions, and is an archetypal feature of rites of passage. But dying and being born again is more than a myth or symbol: it is an actual experience for people who have had a near-death experience.

I think that John the Baptist was a drowner. A person being baptized in the Jordan River, if held under water just long enough, would have had a near-death experience through drowning. In other words, baptism by total immersion could have been a simple, rapid, and effective way of deliberately inducing a near-death experience. Dying and being born again would have been a life-changing personal experience. John may sometimes have gone too far, and some people may not have come back. But that was before liability litigation.

We need myths, and science creates its own. The Big Bang theory, for example, is a version of the ancient creation myth of the hatching of the cosmic egg. As the primal egg cracked open, the universe emerged from it, just as it emerged from the primal singularity of the Big Bang.

I agree with you that one of the major predictors of people's religion is where they were born and the family and culture they were born into. The same goes for atheists. In communist countries, children were indoctrinated into atheism, and many believed what they were taught. In the Soviet Union, the state-sponsored League of the

Militant Godless had 5.5 million members by 1932, and campaigned for the closure of churches, and for their bells to be melted down for industrialization. Parents were warned, "Religion is poison, protect your children!" Such propaganda campaigns were very effective. By 1940, 25 regions of the Soviet Union were declared completely churchless, and fewer than 1,000 churches, chapels, and monasteries survived, compared with 54,000 at the time of the Revolution in 1917.[1] In East Germany, where the communist state vigorously promoted atheism from the 1940s to the 1980s, an NORC University of Chicago survey—"'Beliefs' About 'God' Across 'Time' and 'Countries,'" conducted in 2008, 19 years after the fall of the Berlin Wall—showed that 52 percent of the population were atheists, compared with 10 percent in West Germany. Clearly, people educated to be atheists are more likely to become atheists than those with a non-atheist education.

No doubt some Christians, Muslims, Hindus—and atheists—think that theirs is the One True Faith, but in practice the vast majority do not. I lived for seven years in India, where Hindus, Muslims, Jains, Parsees, Sikhs, Christians, Buddhists, and tribal animists coexist and have done for centuries. Most Indians simply accept that different people follow different religious paths, without feeling the need to convert everyone else to their own religion. Likewise, in North America and in Europe, many religions coexist, as do many Christian churches. Some are bigoted, but most are not. In Britain, where I live, I hardly ever encounter religious fanatics, although I sometimes come across zealous atheists.

Most people who follow a religious path accept that other people have different paths, just as speakers of

English, Turkish, Tamil, or Mandarin accept that other people speak different languages. They do not believe that other languages are false. The existence of different religions does not refute religion in general, any more than the existence of different languages refutes language in general.

You rightly point out that religions serve important social functions, including cooperation between members of a community. But this does not prove that they are nothing but human inventions. Religions have arisen not through philosophical arguments or priestly deceptions, but because many people experience a consciousness greater than their own.

Moreover, the fact that some features of religion are found in many different cultures does not mean, as you suggest, that "we can presume that there is a genetic predisposition for these traits to be expressed within their respective cultures." In his book, *Breaking the Spell* (Viking, 2006), Daniel Dennett proposed a similar and more detailed hypothesis. He proposed that when people in ancient cultures were sick and went to see shamans for a cure, the most credulous would have had the biggest placebo effect and therefore would have tended to survive more. Hence, natural selection would favor genes for credulity and they would be passed on to subsequent generations, increasing the frequency of credulity genes in the population, thus promoting religious belief. But all this is evidence-free speculation. No credulity genes have been found.

Finally, we disagree about gods and God. I think you wrongly conflate them. In polytheistic societies, gods have specific roles, like the god or goddess of the sun, the god or goddess of the sea, and so on. No one imagines that any one of these gods or goddesses or spirits is the ultimate supreme reality, precisely because there are many of them. I agree with you that they are personifications or cultural interpretations of different powers of nature, or of archetypal patterns of experience. But there is a difference between the multiple powers represented by gods, goddesses, spirits, or angels, and the unified source of all things. Even in the most polytheistic of contemporary religions, Hinduism, many Hindus agree that there is a unified source, a supreme conscious reality, one Indian name for which is *satchitananda*, being-consciousness-bliss.

The unified source and sustainer of all things is not just "one god further," but is more fundamental than the many gods and spirits recognized in different religious traditions. Energy provides a scientific parallel. There are many kinds of energy in the universe: the energy in fire, in glucose molecules, in light, in falling rocks, in lithium batteries, in flashes of lightning, and so on. They are all derived from the fundamental energy of the universe, which can take many forms, including the dark energy that propels the expansion of the universe and the evolution of the cosmos. From a religious point of view, this universal energy is a manifestation of divine power. In the Christian tradition, it comes from and is sustained by the power of God. In the Indian tradition, it is called *shakti*.

Materialists share with religious people a belief in a unity underlying all the phenomena of nature. The difference is that religious people believe this ultimate unity is conscious; atheists believe it is unconscious.

—*Rupert*

* * *

Michael Shermer Replies

Dear Rupert,

For my final epistle I would like to pull back and consider the bigger picture across both space and time. We are in the middle of the 2,015th year of the Common Era, a sliver of time that amounts to a mere one-150,000th of the lifespan of our species, one-10,000th of the epoch of our civilization, one-500th of the Age of Science, and one-100th of the Age of Einstein, who discovered that space and time are indivisible. To think that you or I or any of the other seven billion people alive today, or anyone among the 100 billion people who lived before us, has (or had) enough knowledge to know where the universe came from, how life began, the nature of consciousness, the existence of God, the afterlife, and the soul, or what the future holds for humanity, would be *hubris* enough to make a Greek god blush.

Presuming that our species does not go extinct anytime soon through weapons of mass destruction, global climate change, overpopulation, pandemics, a nanotech grey-goo plague, an evil AI or ET, a super–volcano eruption, or a rogue meteor strike on the order of the stone that killed the dinosaurs, we can project ourselves into the future by an amount comparable to these past milestones and imagine what people will know about these great mysteries a century from now in the year 2115, or in half a millennium in the year 2515, in 10 thousand years in 12015, or 150,000 years from now in the year 152015. Unimaginable. Literally. Given the current accelerating rate of change in which the world has transformed more in the past century than it did in the previous 10 centuries, and will change in the 21st century more than it did in the previous 100 centuries, it is impossible to know what people "in the year 2525" will know (to quote a once popular song), much less our descendants in the year 9595—the *exordium et terminus* of this civilizational journey.

A brief survey of the history of science alone should humble us into acknowledging that the answers to these existential questions may not be forthcoming anytime soon, and that our current best theories—as well supported as some of them are—may one day go the way of Aristotelian physics, Ptolemaic astronomy, the flat-earth theory, the hollow-earth theory, the expanding-earth theory, phlogiston theory, the miasma theory of disease, the four bodily humors theory of medicine, phrenology, preformationism, creationism, alchemy, astrology, the luminiferous aether, the Rutherford model of the atom, the steady-state theory of the cosmos, and vitalism. Among the many good reasons to want to live a long and healthy life, or even be cryonically frozen and brought back to life centuries from now, would be to find out what we were wrong about in the 20th and 21st centuries. As we look back on these erroneous scientific theories from centuries past with disdainful dismissal, what will scientists in the 26th—or the 260th—century think about our current models of the cosmos, life, and consciousness? Will they look down upon us as we do medieval physicians who believed that the four bodily humors (black bile, phlegm, blood, and yellow bile) were linked to the four elements of nature (earth, water, air, and fire) that caused the four personality temperaments (melancholic, phlegmatic, sanguine, and choleric), all of which were linked to the zodiacal signs in the heavens?

That said, when I am ill I opt for the best medicine available in 2015, not 1515, because we really have learned something about the human body over the past half millennium. The same assumption of cumulative progress in science applies to cosmology, physics, biology, and psychology—our current models really are better than our past models, and the fact that some of these theories were wrong does not mean all current ones are mistaken (and thus every alternative theory must be taken seriously). Your alternative theories about the human mind and consciousness may in centuries hence turn out to be right, but the odds are long against it.

Still, as I document in *The Moral Arc*, one of the greatest discoveries of the Scientific Revolution and the Enlightenment is that free expression and the open marketplace of ideas where everyone is welcome to proffer their beliefs without state censorship has been one of the driving forces behind both scientific and human progress. I'm skeptical of your theories, Rupert, but I defend your right to publish and present them in public forums such as this, where they can be exposed to the bright light of science—the most important invention ever made in the history of our species.

—*Michael*

Footnotes

Notes for Rupert Sheldrake Interview

1. Johann Wolfgang von Goethe (1749–1832), the author of *The Sorrows of Young Werther*, *Elective Affinities*, *Wilhelm Meister*, *Poetry and Truth*, *East-West Divan*, *Faust* (Parts I and II), and many other works in prose and verse. Goethe was very interested in the natural sciences, conducting his own observations and experiments. He published many works on scientific subjects, including *The Metamorphosis of Plants* (1790) and *Theory of Color* (1810), as well as numerous shorter scientific studies.

2. University of Chicago Press, 1962.

3. See Donna J. Haraway, *Crystals, Fabrics, and Fields: Metaphors of Organicism in Twentieth-Century Developmental Biology*, Yale University Press, 1976; and Erik Peterson, "The Conquest of Vitalism or the Eclipse of Organicism? The 1930s Cambridge Organizer Project and the Social Network of Mid-Twentieth-Century Biology," British Journal for the History of Science, 2014, 47: 281–304.

4. Univeristy of Michigan Library, 1896.

5. Blond & Briggs, 1981. First U.S. ed.: J.P. Tarcher, 1982; second U.S. ed.: Park Street Press, 1999; third U.S. ed. (retitled *Morphic Resonance: The Nature of Formative Causation*): Park Street Press, 2009.

6. Deepak Chopra, 2013; p. 93.

7. Oxford University Press, 2012; p. 1

8. 'Scientism' is a term that occurs repeatedly throughout this text. The authors disagree as to its fundamental definition, and in fact, their disagreement about this singular word's definition goes to the heart of their disagreement about the nature of science. According to Dr. Shermer: scientism is a scientific world view that encompasses natural explanations for all phenomena, eschews supernatural and paranormal speculations, and embraces empiricism and reason as the twin pillars of a philosophy of life appropriate for an Age of Science. According to both Dr. Sheldrake and the Merriam Webster Dictionary, scientism is [an] "exaggerated trust in the efficacy of the methods of natural science applied to all areas of investigation (as in philosophy, the social sciences, and the humanities)."

Notes for Michael Shermer Interview

1. Henry Holt and Co., 2015. The title alludes to a famous line from one of Dr. Martin Luther King, Jr.'s, speeches: "The arc of the moral universe is long, but it bends towards justice."
2. "Rupert's Resonance," *Scientific American*, October 24, 2005.
3. The exact relationship between the earlier "natural law" and the later "natural rights" traditions is convoluted and, in part, contested; however, there is no doubt of the deep connection between the two traditions. For the philosophical issues involved, see John Finnis, *Natural Law and Natural Rights*, Oxford UP, 1980. For the history, see M.B. Crowe, *The Changing Profile of the Natural Law*, Martinus Nijhoff, 1977, Knud Haakonssen,

Natural Law and Moral Philosophy: From Grotius to the Scottish Enlightenment, Cambridge UP, 1996, and Brian Tierney, *The Idea of Natural Rights: Studies on Natural Rights, Natural Law, and Church Law*, (1150–1625), Scholars Press, 1997.

4. Viking, 2011.

5. Joseph Butler, *Five Sermons Preached at the Rolls Chapel and a Dissertation upon the Nature of Virtue*, ed. by Stephen L. Darwall, Hackett Publishing Co., 1983; originally published as a group of 15 sermons in 1726. See, especially, the first three sermons, which are collectively entitled "Upon Human Nature." This idea and some of its history have been beautifully developed in a wonderful little book called *The Inner Check*, E. Wright, 1974, by the late Swedish philosopher Folke Leander, which we recommend to everyone.

Notes to Shermer's Opening Statement on 3a. Mental Action at a Distance

1. Wiseman, Richard and Marilyn Schlitz (1997) "Experimenter Effects and the Remote Detection of Staring," *Journal of Parapsychology* **61**: 197–207.

2. Freeman, Anthony, ed. (2005) "Sheldrake and His Critics: The Sense of Being Glared At," *Journal of Consciousness Studies*, **12**(6).

3. 115: 4–18.

4. Hyman, Ray (1994) "Anomaly or Artifact? Comments on Bem and Honorton,"*Psychological Bulletin*, **115**: 19–24.

6. Milton, Julie and Richard Wiseman (1999) "Does Psi Exist? Lack of Replication of an Anomalous Process

of Information Transfer," *Psychological Bulletin*, 125: 387–391.

Notes to Rupert Sheldrake's Response on 3b.
Mental Action at a Distance

1. Williams, B.J. (2011) "Revisiting the Ganzfeld ESP Debate: A Basic Review and Assessment," *Journal of Scientific Exploration*, **25**: 639–661 (Download PDF: DeanRadin.com/evidence/Williams2011Ganz.pdf) 2. Milton, J. (1999) "Should ganzfeld research continue to be crucial in the search for a replicable psi effect?" *Journal of Parapsychology*, **63**: 309–333. 3. Mossbridge, et al. (2012) "Predictive physiological anticipation preceding seemingly unpredictable stimuli: a meta-analysis," *Frontiers in Psychology* **3**: 1–16 (Download PDF: DeanRadin. com/evidence/Mossbridge2012Presentiment.pdf) 4. Carter, C. (2010) "'Heads I lose, Tails you win,' or, How Richard Wiseman nullifies positive results and what to do about it," *Journal of the Society for Psychical Research*, **74**: 156–167; McLuhan, R. *Randi's Prize: What Skeptics Say About the Paranormal, Why They Are Wrong and Why It Matters*. Matador, 2010. Storr, W. *The Unpersuadables: Adventures with the Enemies of Science*. Overlook Press, 2014. 5. Watt, C., R. Wiseman, and M. Schlitz (2002) "Tacit information in remote staring research: The Wiseman-Schlitz interviews," *Paranormal Review* **24**: 18–25.

Notes to Rupert Sheldrake's Opening Statement on
5A. God and Science

1. For an illuminating discussion of the traditional understanding of God in the Christian and other religious traditions, see David Bentley Hart, *The Experience of God: Being, Consciousness, Bliss.* Yale University Press, 2013.
2. New Revised Standard Version (NRSV).
3. Alistair Hardy, *The Spiritual Nature of Man.* Oxford: Oxford University Press, 1979.
4. Harold G. Koenig, *Medicine, Religion and Health: Where Science and Spirituality Meet.* Templeton Foundation Press, 2008.
5. *Ibid.*; p. 143.

A Note to Rupert Sheldrake's Response on 6B.
God and Science

1. Nick Spencer, *Atheists: The Origin of the Species.* London: Bloomsbury Academic, 2014; Chap. 4.